Robust Stormwater Management in the Pittsburgh Region

A Pilot Study

Jordan R. Fischbach, Kyle Siler-Evans, Devin Tierney, Michael T. Wilson, Lauren M. Cook, Linnea Warren May

For more information on this publication, visit www.rand.org/t/RR1673

Library of Congress Cataloging-in-Publication Data is available for this publication.
ISBN: 978-0-8330-9795-8

Published by the RAND Corporation, Santa Monica, Calif.
© Copyright 2017 RAND Corporation
RAND® is a registered trademark.

Cover: GettyImages/sdominick

Support RAND
Make a tax-deductible charitable contribution at
www.rand.org/giving/contribute

www.rand.org

Preface

Cities and larger metropolitan regions are at the forefront of efforts to reduce greenhouse gas emissions, understand and respond to current and future effects of climate change, and develop resilience and adaptation capacity in response to climate change. Through the support of the John D. and Catherine T. MacArthur Foundation, the RAND Corporation is helping advance these efforts by bringing new analytical and planning capabilities pioneered at state and multistate levels into a regional urban context. This effort includes pilot studies in three urban areas.

The pilot studies have several overall objectives: (1) to add value to ongoing deliberative processes surrounding challenging investment, policy, or other climate-related problems in the region; (2) to identify and analyze a range of potential solutions in an open and interactive public planning process; (3) to facilitate consensus around an effective and fiscally sustainable approach; and (4) to build in-house capacity within regional organizations to tackle these increasingly complex planning choices on their own using best available analytical methods. Experience with these pilot studies will also inform the foundation's deliberations on its future investment decisions.

This study focused on the ongoing challenge of stormwater management in the Pittsburgh metropolitan region. The City of Pittsburgh and other municipalities in Allegheny County, Pennsylvania, face significant challenges in meeting water-quality requirements and upgrading their aging and inadequately sized regional combined sewer system, a problem that could grow with future climate, population, or land-use changes. This report provides an independent study of the growing stormwater problem in the Pittsburgh, Pennsylvania, metropolitan region and discusses potential long-term solutions using new analytical approaches developed by RAND in other contexts. The intended audience includes local government agencies and regional authorities addressing this challenge, local stakeholders engaged in stormwater and wastewater planning, state and federal regulators, and planners in other cities facing similar challenges with aging infrastructure and climate uncertainty.

Interested readers should also see the following RAND publication that sets forth the broader planning framework for this effort: Debra Knopman and Robert J. Lempert, *Urban Responses to Climate Change: Framework for Decisionmaking and Supporting Indicators*, Santa Monica, Calif.: RAND Corporation, RR-1144-MCF, 2016.

RAND Infrastructure Resilience and Environmental Policy

The research reported here was conducted in the RAND Infrastructure Resilience and Environmental Policy program, which performs analyses on urbanization and other stresses. This includes research on infrastructure development; infrastructure financing; energy policy; urban planning and the role of public-private partnerships; transportation policy; climate response, mitigation, and adaptation; environmental sustainability; and water resource management and coastal protection. Program research is supported by government agencies, foundations, and the private sector.

This program is part of RAND Justice, Infrastructure, and Environment, a division of the RAND Corporation dedicated to improving policy- and decisionmaking in a wide range of policy domains, including civil and criminal justice, infrastructure protection and homeland security, transportation and energy policy, and environmental and natural resource policy.

Questions or comments about this report should be sent to the project leader, Jordan Fischbach (jordan_fischbach@rand.org). For more information about RAND Infrastructure Resilience and Environmental Policy, see www.rand.org/jie/irep or contact the director at irep@rand.org.

Contents

Figures

Tables

Summary

Pennsylvania's Allegheny County faces a monumental water management challenge. Eighty-three of the 130 municipalities in the county, including the City of Pittsburgh, rely on the Allegheny County Sanitary Authority (ALCOSAN) system to manage regional stormwater and wastewater flows. This sewer system consists of thousands of miles of pipes, many of which are leaking or broken, that connect to a single treatment plant operated by the authority. ALCOSAN directly manages only a small portion of this system, with individual municipalities and authorities responsible for operating and maintaining most of the extensive pipe network.

The system is inadequately sized to capture and treat most "wet weather" events (rain and snowfall), which occur frequently throughout the year. As a result, nearly every time it rains, a sewer overflow occurs in at least one of the approximately 450 outfalls somewhere in the system, draining a mix of untreated wastewater and stormwater into the county's streams and rivers (Figure S.1). A mixture of wastewater and stormwater entering waterways is called a *combined sewer overflow* (CSO). In some places, undiluted sewage can flow directly from the wastewater system into nearby water bodies, referred to as a *sanitary sewer overflow* (SSO). More than 9 billion gallons (Bgal.) of sewer overflow in a typical year have led to violations of the U.S. Clean Water Act (33 U.S.C. 1251), along with state and county public health laws.

Adding to the challenge is potential deep uncertainty as to how the overflow problem might change in the future. Recent projections suggest that the northeastern United States could see increased precipitation caused by climate change, which challenges the traditional practice of using historical observations alone as a basis for future water infrastructure planning and design. The planning challenge could be further exacerbated by population growth and changing land-use practices. Decisionmakers and planners in Allegheny County communities must make crucial infrastructure and policy decisions to improve the current situation and satisfy regulatory requirements, but currently lack information about future changes. These decisions will have implications for many decades to come, so the choices made today must also be resilient to possible changes and future conditions.

This pilot study has taken a first step toward addressing these planning needs. It provided a baseline of scientific information regarding future uncertainty to support

Figure S.1
ALCOSAN System Map

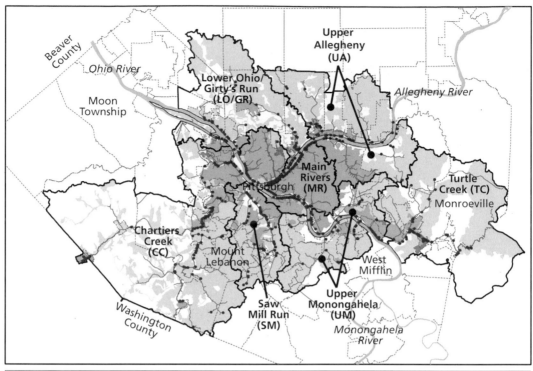

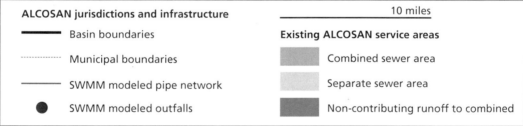

SOURCE: Data from Allegheny County Sanitary Authority.
NOTE: SWMM = Storm Water Management Model.

RAND *RR1673-S.1*

ongoing regional coordination around stormwater management and wet weather planning for municipalities in the ALCOSAN sewer system. The study draws on RAND's analytical, modeling, and planning capabilities and is one of three pilots to use a new framework for urban adaptation and response to climate change. We included the Pittsburgh metropolitan region as a pilot because its challenges are similar to those in many other older urban areas in the United States. Most of these areas are struggling to rebuild and modernize aging infrastructure and work across jurisdictional lines to solve regional water management problems. All are facing a changing climate, economic transformations, and shifts in land use.

The Infrastructure Planning Challenge Remains Unresolved

ALCOSAN estimates that, under historical rainfall conditions, approximately 9.5 Bgal. of CSOs and SSOs can occur each year. Given the negative effects on water quality, ecosystems, and public health from these overflows, the U.S. Environmental Protection Agency in 2008, working with its state and local partners, issued a consent decree requiring ALCOSAN and its municipal customers to meet water-quality requirements mandated by the U.S. Clean Water Act and state and county laws. To meet this legal mandate, ALCOSAN is required to make substantial improvements to the sewer system by 2026 to eliminate SSOs and greatly reduce the frequency and volume of CSOs during wet weather events. However, because ALCOSAN owns or directly manages only a small portion of the sewer system, an effective solution will require coordination and partnership with the 83 contributing municipalities, including the City of Pittsburgh.

In response to the consent decree, ALCOSAN developed a draft Wet Weather Plan (WWP) in 2012. The WWP was initially designed using traditional wastewater management infrastructure (e.g., treatment, pipes, tunnels, and storage tanks), sometimes referred to as "gray" infrastructure. Notably, the WWP includes an expansion of the wastewater treatment plant and a series of new deep tunnels intended to carry away and store combined stormwater and wastewater during rainfall events. These proposed investments, with a total capital cost of approximately $2 billion, are intended to store and treat a substantial new volume of water to significantly reduce SSOs and CSOs by 2026. The draft WWP has not yet been finalized or accepted by the regulators, however, and the key parties continue to negotiate the plan's content and its implementation schedule.

One key point of negotiation is the proposed inclusion of "source reduction"—that is, investments to reduce or prevent groundwater or stormwater from entering the existing sewer system "at the source," thereby freeing up capacity and reducing overflows. The proposed source reduction approaches for ALCOSAN communities include repairs to existing sewer pipes and such "green" approaches as green stormwater infrastructure (GSI). GSI is intended to divert and capture water either temporarily (with eventual flow back to the system) or permanently (through infiltration into the ground, evapotranspiration, or another path of flow to the rivers) using vegetation, soils, or other natural elements, rather than allow it to flow directly into the sewer system during a storm. The City of Pittsburgh and other local stakeholders support the broader use of source reduction and GSI to augment or offset portions of the WWP. However, information is lacking on how GSI options perform at scale and how they compare with the performance of traditional gray wastewater management infrastructure.

This Study Uses a Participatory Approach and Robust Decision Making Methods

The study is based on a participatory decision support framework and method for improving decisions under deep uncertainty called Robust Decision Making (RDM). RDM is an iterative, quantitative decision analytic framework that brings together experts and decisionmakers to help identify the full extent of a challenge, as well as potentially robust strategies to address it. The framework then enables the participants to characterize the vulnerabilities of the proposed strategies and evaluate the trade-offs among them.

Two groups of regional stakeholders informed the study. The first consisted of *Study Partners*, who included planners and technical experts already conducting relevant research and analysis at the county or city level. Study Partners provided data, models, and/or modeling support, helped to identify and assess options, and assisted in the estimation of associated project-level and strategy costs. The second group, *Stakeholder Advisors*, included decisionmakers from some local municipalities, along with representatives from different economic development organizations, watershed associations, non-government organizations, and community groups. This group served to ensure the accuracy, reliability, and relevance of the emerging analysis. Both groups participated in the deliberation process.

For the first stage of the process, *scoping*, Study Partners and Stakeholder Advisors were asked to help define the research scope and identify the inputs needed to conduct the technical analysis. Specifically, participants helped define the goals to be met and associated performance metrics used to quantify these goals (M), policy levers or broader strategies that could be implemented to achieve these goals (L), uncertain factors that could affect the ability to achieve these goals but were outside the control of decisionmakers (X), and relationships among these elements as reflected in simulation or planning models (R). Summary results from these discussions are shown in Table S.1, with each quadrant representing one of these key elements.

Participants in the scoping workshops identified a wide-ranging and ambitious research agenda, which could be considered the scope of an integrated regional watershed management planning effort. Drawing from this broad scope, the research team identified a set of discrete analytic steps that could be accomplished within the time frame and resources of this pilot study.

Analysis of Future Vulnerability

Next, a scenario analysis approach to explore future sewer system vulnerability was applied. Using a set of existing simulation models developed by ALCOSAN, the RAND team developed a new automated simulation framework and conducted a

Table S.1
Scoping Summary from Partner and Stakeholder Workshops

Uncertain Factors (X)	Policy Levers and Strategies (L)
• **Climate change** • **Land-use, development, or environmental changes** • **Economy and population changes** • **Demand for water or wastewater services** • **Cost-effectiveness and affordability** • Regulatory or political landscape • Popular opinions or public sentiment • Stormwater or wastewater modeling uncertainty • **Infrastructure performance uncertainty**	• **Gray infrastructure (tunnels, pipes, treatment, storage)** • **Stormwater source reduction** • **Retrofitting, repair, operations, and maintenance** • **GSI** • **Regulations on land use and zoning** • Integrated ecosystem services • Centralized management organization or plan • Market-based solutions, innovative incentive design, or financing • Education, strategic communication, or public awareness

Relationships (R)	Goals and Metrics (M)
• **Hydrologic and hydraulic models** • Flood risk models • **Downscaled climate-informed hydrology** • **Land-use change model** • **Infrastructure cost estimation tools**	• **Improve water quality** • **Reduce sewer overflow** • Reduce flood risk • **Comply with regulatory requirements** • **Protect infrastructure** • Protect and improve ecosystem • Improve property values, add community amenities, and reduce risk premium • **Build regional cooperation and coordination** • Gain public support

NOTE: Responses in bold represent those that the RAND team was able to carry forward into the technical analysis.

series of quantitative experiments using high-performance cloud computing to explore how the system might respond to plausible future scenarios reflecting the number of future wastewater customer connections, land-use changes, or climate change. All of these results are based on a "future without action," in which the current system is maintained as it is; in addition to stress-testing the existing system for vulnerabilities, this approach provides a baseline for comparing future investments.

In this analysis, two different assumptions about wastewater customer connections, three assumptions about land use, and three assumptions about climate change were developed and considered. The land-use and climate change assumptions were developed as part of this study, requiring new research and analysis.

- **Climate change:** Expanding on previous work that used rainfall from a single average or "typical" year (2003 Typical Year), the team created a recent historical rainfall scenario using observed data from 2004 through 2013 (Recent Historical), and then developed two climate-adjusted rainfall and temperature scenarios using projections from 2038 through 2047. Higher Intensity Rainfall represents a scenario with the highest-intensity daily storms, yet a marginal increase in total annual rainfall, while Higher Total Rainfall projects the largest increase in total annual rainfall and slightly less intense daily storms.

- **Land use:** Three land-use scenarios were developed reflecting no population growth (Current Land Use), moderate growth (Southwestern Pennsylvania Commission [SPC] Growth), and high growth (2xPGH). These scenarios change the amount of impervious cover (pavement or buildings) present in the areas of the sewer system contributing to CSOs. Impervious cover prevents the infiltration of rain into the soil where it first falls, leading to more overland flow and less infiltration into the groundwater. In this analysis, future land use is treated as an uncertainty, and land-use regulations are not modeled as policy levers. However, enforcement of existing or new land-use regulations could help to limit or avoid increases in impervious cover, reducing the likelihood of such scenarios as SPC Growth or 2xPGH.

- **Wastewater customer connections:** Adapting results from the ALCOSAN analysis, the vulnerability analysis includes a scenario representing the current number of customer connections (Current Connections), as well as one with increased inflows resulting from an increase in the number of customer connections and an expansion of the ALCOSAN service area (Future Connections). The Future Connections scenario adds approximately 500 million gallons (Mgal.) per year of overflow in this analysis.

Each unique combination of assumptions across the uncertain factors is described as a "scenario" in this report. All possible combinations were considered, yielding a total of 18 scenarios simulated for the vulnerability analysis.

Figure S.2 summarizes the results of this exercise, showing total sewer overflows at designated outfalls simulated for the region in each scenario. Compared with a simulation of the 2003 Typical Year that matches recent ALCOSAN estimates, this figure shows the extent to which overflows could increase.

This analysis produced the following key findings:

- **Simulations of the recent past suggest that overflow volumes are up to 15 percent higher than previously estimated.** Results from the Recent Historical climate scenario show that the overflow challenge may have already grown in the past decade, with sewer overflows increasing from approximately 9.5 Bgal./ year in a 2003 Typical Year simulation to a ten-year average of 11 Bgal./year for 2004 to 2013 when holding other system characteristics constant. In part, these increases could be because of an increase in the average annual rainfall, but likely also reflect differences in storm patterns and intensity when comparing the recent ten-year period with the 2003 Typical Year.

Figure S.2
Overflow Results from All Overflow Scenarios Considered

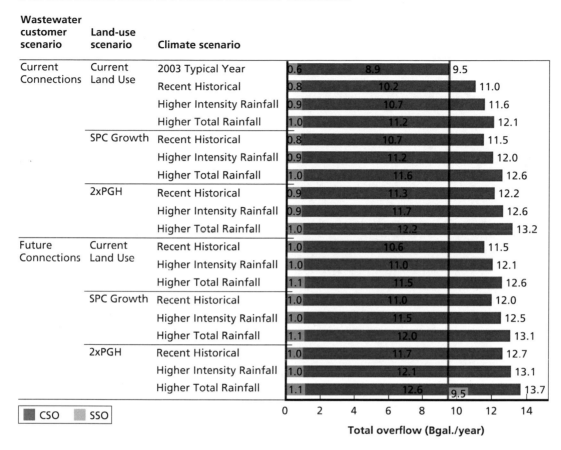

Wastewater customer scenario	Land-use scenario	Climate scenario		
Current Connections	Current Land Use	2003 Typical Year	0.6	8.9	9.5
		Recent Historical	0.8	10.2	11.0
		Higher Intensity Rainfall	0.9	10.7	11.6
		Higher Total Rainfall	1.0	11.2	12.1
	SPC Growth	Recent Historical	0.8	10.7	11.5
		Higher Intensity Rainfall	0.9	11.2	12.0
		Higher Total Rainfall	1.0	11.6	12.6
	2xPGH	Recent Historical	0.9	11.3	12.2
		Higher Intensity Rainfall	0.9	11.7	12.6
		Higher Total Rainfall	1.0	12.2	13.2
Future Connections	Current Land Use	Recent Historical	1.0	10.6	11.5
		Higher Intensity Rainfall	1.0	11.0	12.1
		Higher Total Rainfall	1.1	11.5	12.6
	SPC Growth	Recent Historical	1.0	11.0	12.0
		Higher Intensity Rainfall	1.0	11.5	12.5
		Higher Total Rainfall	1.1	12.0	13.1
	2xPGH	Recent Historical	1.0	11.7	12.7
		Higher Intensity Rainfall	1.0	12.1	13.1
		Higher Total Rainfall	1.1	12.6	9.5 13.7

CSO   SSO

Total overflow (Bgal./year)

0   2   4   6   8   10   12   14

NOTES: Light and dark gray in the stacked bar chart show the separate contributions of SSOs and CSOs, respectively, to total overflow volume. The vertical gray line shows the estimated total overflow using Current Connections, Current Land Use, and 2003 Typical Year rainfall for visual comparison purposes.

RAND RR1673-S.2

- **Future rainfall, population, and land-use changes could increase overflow volumes.** This exercise showed that overflows could grow further with plausible future changes. Figure S.2 shows that all uncertain factors may contribute to overflow increases but that no single factor is dominant. The extent of increased vulnerability depends on the assumptions, but it ranges from 1.5 to 4.2 Bgal./year in additional overflow volume (11 to 13.7 Bgal./year in total) when compared with a 2003 Typical Year simulation. Of this total, SSOs increased to between 0.8 and 1.1 Bgal./year in the future simulations with no additional policy action.

Strategies to Reduce Future Overflow

Building on the vulnerability analysis, the RAND team next evaluated a set of regional, planning-level source reduction strategies for the ALCOSAN service area. Participants in the process helped identify policy levers for this analysis intended to either improve the function of the existing sewer system or reduce the flow of stormwater into the system during rainfall events.

Specifically, a "no action" baseline was considered, along with four other lever types:

- **No action:** This represents the sewer system as presently constructed and operated with no additional investments or improvements made.
- **GSI:** This lever encompasses a wide range of technologies and approaches for managing stormwater runoff, including rain barrels, rain gardens, bioretention, infiltration trenches, and green roofs. This analysis used a simplified and high-level approach to evaluate GSI in the simulation modeling, targeting the control of stormwater runoff from 10 to 40 percent of impervious cover in the combined sewer area with different design assumptions. Table S.2 summarizes the names and assumptions used for each GSI strategy.
- **Inflow and infiltration (I&I) reduction:** I&I is the result of an aging sewer system with leaks in manholes and pipes, as well as some cases in which buried streams flow directly into the system. This lever includes pipe repair approaches aimed at reducing rainfall-derived I&I, which occurs during and after rainstorms, and groundwater inflows, which can occur in wet or dry periods. The team developed six possible I&I strategies based on three levels of inflow reduction that

Table S.2
GSI Strategy Assumptions

Strategy Name	GSI Type	GSI Sizing	Infiltration Rate (inches per hour)
GSI-10	Bioretention	10% of DCIA with 1" rain	0.1
GSI-20	Bioretention	20% of DCIA with 1" rain	0.1
GSI-40	Bioretention	40% of DCIA with 1" rain	0.1
GSI-40-HI (High Infiltration)	Bioretention	40% of DCIA with 1" rain	0.2
GSI-40-AOP (Art of the Possible)	Bioretention	40% of DCIA with 1.5" rain	0.2

NOTE: Strategy names refer to the percentage of directly connected impervious area (DCIA) used to determine the size of the GSI. For example, GSI-20 is sized according to the volume of runoff from 1 inch of rainfall over 20 percent of DCIA in the combined service area.

could be achieved (low, mid, or high) and two different target areas (areas with inflow or R-values greater than 6 percent or 8 percent).

- **Treatment plant expansion:** This refers to expanding ALCOSAN's Woods Run wastewater treatment plant to a capacity of 480 Mgal. per day (MGD), up from the current capacity of 250 MGD.
- **Deep-tunnel interceptor cleaning:** This refers to cleaning the existing main interceptor tunnels along the Allegheny, Monongahela, and Ohio rivers to increase conveyance and storage capacity.

Next, the simulation modeling framework was employed to conduct a preliminary screening analysis of 30 different strategies, which included an evaluation of each of these levers in isolation or in combination. In the screening, each strategy was evaluated in one scenario only, with a single assumption about rainfall (2003 Typical Year), wastewater customer connections (current), land use (current), and capital cost ("nominal" or mid-range estimate for each lever type). The goal of this screening analysis was to provide preliminary insights on the relative performance of different policy levers using a relatively small number of simulation runs, with results that could be compared with those from recent investigations by ALCOSAN and other organizations. It was also intended to identify a promising subset of strategies to consider in the more complete RDM uncertainty analysis.

All screening strategies were evaluated relative to a baseline of 9.5 Bgal./year in total overflow. With varying design and effectiveness assumptions for each lever type, simulation results showed that the treatment expansion alone reduced overflows by 25 percent, while treatment expansion together with cleaning of the main interceptors reduced overflows by 34 percent. I&I strategies alone reduced total overflow 5 to 19 percent, and GSI strategies reduced overflows 5 to 21 percent. This screening also tested a variety of combination strategies—which included a mix of the different lever types—yielding overflow reductions of 35 to 62 percent.

RDM Analysis

RDM techniques were then used to evaluate a subset of ten promising strategies identified in the screening analysis across a range of nearly 5,000 uncertain scenarios. All but one of the strategies evaluated in this phase included a treatment plant expansion, along with one or more additional policy lever types. Uncertain factors considered include climate, wastewater customer connections, and land use, as described earlier. Additional uncertain factors were also included in this phase, such as cost uncertainty for each lever type and GSI performance uncertainty (Low or High performance, with differing assumptions about how much stormwater runoff can be routed to GSI project sites).

Overflow results for the selected strategies across all uncertain scenarios are summarized in Figure S.3. Each box plot shows the range of total overflow (Bgal./year) with that strategy implemented across all uncertain scenarios. The first row in the figure shows the range of results in a case of future without action (no action). The selected strategies simulated improve on these results: For example, implementing GSI-20 alone (Strategy 2) reduces total overflow to 8.2 to 12.5 Bgal./year, a reduction of 0.7 to 2.1 Bgal./year. Upgrading the treatment plant (Strategy 12) instead reduces total overflow to 7.0 to 10.4 Bgal./year (2.5- to 3.6-Bgal./year reduction).

In general, including additional system improvements or combining policy levers further reduces overflow relative to a future without action. Coupling treatment plant expansion with GSI-20 (Strategy 15), for example, reduces overflow by 3.3 to 5.3 Bgal./year but leaves 5.8 to 9.1 Bgal./year of overflow remaining. The final four strategies

Figure S.3
Remaining Overflow with Selected Strategies, All Scenarios

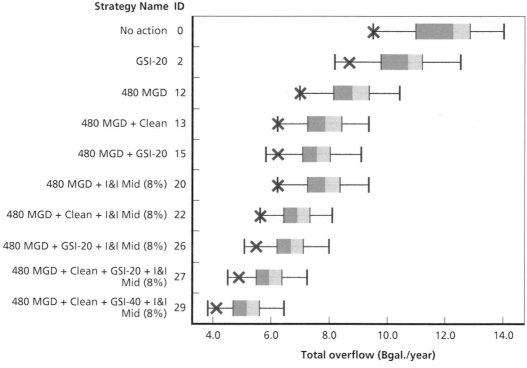

NOTE: The box plots presented do not represent probability distributions but instead report the results of a set of model runs (scenarios). Strategies with GSI are evaluated in 48 unique scenarios, and those without GSI show 24 scenario results. Each point summarized represents one mapping of assumptions to consequence, and the points are not assumed to be equally likely. Components of the box plot include the 25th and 75th percentiles (edges of each box), median (vertical line where the two gray shaded areas meet), and extremes of the data set (whiskers). The blue X indicates an initial set of assumptions used in strategy screening that are similar to those applied in other recent research. 480 MGD = treatment plant expansion; Clean = deep-tunnel interceptor cleaning.
RAND RR1673-S.3

considered include a combination of three or more levers. These strategies generally yield the most total overflow reduction. They also reduce SSOs by approximately 250 to 470 Mgal./year but leave 300 to 600 Mgal./year remaining (not shown).

These results suggest that source reduction, when combined with a treatment plant expansion and/or cleaning of the existing deep tunnels, can reduce overflows if strategies perform comparably to the simplified, planning-level assumptions. However, different assumptions about future climate, population, or land use substantially influence strategy performance, yielding wide ranges of uncertainty around the remaining overflows. In addition, it shows that **none of the strategies combining treatment plant expansion with source reduction fully eliminates SSOs or nearly eliminates CSOs overflows in any scenario**. No strategy considered here appears to fully resolve this significant challenge, and more investigation is needed to understand how these approaches could be combined with additional infrastructure investments to reliably eliminate sewer overflows as part of a long-term solution.

Strategy Cost-Effectiveness with Uncertainty

Next, the team incorporated preliminary, first-order estimates of capital cost to calculate the cost-effectiveness of selected strategies across a range of scenarios. This portion of the analysis combined uncertainty results for overflow reduction with an additional set of cost uncertainties for each policy lever type, resulting in an evaluation across nearly 5,000 uncertain scenarios.

Figure S.4 shows a box plot summary of cost-effectiveness results, represented in terms of 2016 dollars per gallon of overflow reduced, for eight strategies. Each box plot summarizes strategy performance across all overflow and capital cost uncertainties (4,944 scenarios). A lower value in this metric is better, implying a lower cost to achieve each additional gallon of annual overflow reduction. A line is included at $0.35 per gallon, which is the average cost-effectiveness estimated by ALCOSAN for the draft WWP (in 2016 dollars) and serves as a convenient reference point for cost-effectiveness comparisons.

Implementing GSI-20 alone (first row) yields a wide range of plausible cost-effectiveness, depending on the overflow and cost scenario assumption, from $0.14 to $1.59 per gallon, with many scenarios above the reference value of $0.35 per gallon. Strategies including treatment plant expansion alone or together with interceptor cleaning, by contrast, yield good cost-effectiveness performance across the full range of scenario uncertainty, with all scenario results below $0.20 or $0.25 per gallon. These results show that **expanding wastewater treatment plant capacity or cleaning existing deep interceptors could represent low-regret, near-term options**.

The remaining strategies, combining treatment or treatment and interceptor cleaning with source reduction, show a more mixed picture of cost-effectiveness. In general,

Figure S.4
Cost-Effectiveness for Selected Strategies, All Scenarios

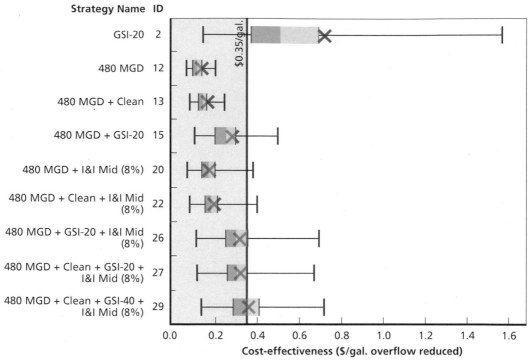

NOTE: The box plots presented do not represent probability distributions but instead report the results of a set of model runs (scenarios). Strategies with GSI are evaluated in 4,944 unique scenarios, and those without GSI show 2,472 scenario results. Each point summarized represents one mapping of assumptions to consequence, and the points are not assumed to be equally likely. Components of the box plot include the 25th and 75th percentiles (edges of each box), median (vertical line where the two gray areas meet), and extremes of the data set (whiskers). The blue X indicates the set of assumptions used in the screening analysis. Cost is represented in 2016 constant dollars, and the red reference line shows the average cost-effectiveness of ALCOSAN's draft WWP. Yellow shading indicates cost-effectiveness below this threshold.
RAND RR1673-S.4

simulation results show that **combining a wastewater treatment plant expansion with incremental investments in source reduction could yield cost-effective overflow reduction under plausible future assumptions**. However, the uncertainty associated with both overflow reduction performance and source reduction costs is generally high, yielding wide ranges in cost-effectiveness performance.

Scenario Discovery to Identify Key Uncertain Drivers

As a final step in the RDM analysis, the RAND team used scenario discovery methods to help identify the key uncertain drivers for cost-effectiveness of either a GSI-only strategy (GSI-20; Strategy 2) or for one possible Combined Source Reduction Strategy

(Strategy 27). Using this process, the RAND team sought to identify uncertain drivers that yield cost-effectiveness either below (acceptable) or above (poor) the threshold of $0.35 per gallon.

Using statistical algorithms and visualization tools, several key drivers were identified common to both strategies: (1) future climate uncertainty (represented as average annual rainfall); and (2) the average capital cost of GSI per impervious acre controlled. A combination of these factors, together with a handful of other drivers, most frequently leads to either acceptable or poor cost-effectiveness for these strategies:

- **Strategy 2 (GSI-20):** Implementing GSI-20 alone reduces total overflow 0.8 to 2.1 Bgal./year across the scenario range, with a cost range of $0.3 billion to $1.1 billion. This strategy leads to acceptable cost-effectiveness in about one-quarter of the simulated scenario results (Figure S.4, left tail). Assuming historical climate assumptions (2003 Typical Year or Recent Historical rainfall), GSI-20 yields acceptable cost-effectiveness only when assuming High GSI performance and low average per-acre capital costs. In plausible future climate scenarios with higher rainfall volumes, however, GSI-20 cost-effectiveness increases. Here, the strategy can be cost-effective assuming Low GSI performance and across a wider range of capital costs overall ($283,000 to $353,000 per acre).

- **Strategy 27 (Combined Source Reduction Strategy):** The Combined Source Reduction Strategy includes the treatment plant expansion, interceptor cleaning, and investments in both I&I and GSI source reduction. It yields total overflow reduction of 4.6 to 7.1 Bgal./year across the range of scenarios considered. Capital costs estimated for this strategy range from $0.8 billon to $3.1 billion, yielding a range of cost-effectiveness from $0.12 per gallon to $0.67 per gallon.

  This strategy is also increasingly cost-effective as the average annual rainfall increases and the climate scenario becomes more adverse. Strategy 27 simulation results show acceptable cost-effectiveness in nearly all scenarios that include plausible future increases in annual rainfall. Under historical climate assumptions, alternatively, poor cost-effectiveness performance emerges only with either higher average GSI per-acre costs (more than $324,000 to $415,000) or a high percentage of pipes needing repairs to meet I&I reduction targets (higher than 76 percent).

The RDM analysis shows that investing in GSI source reduction to control up to 20 percent of impervious cover in the combined sewer system, when evaluated in isolation, yields poor cost-effectiveness for overflow reduction under "nominal" or commonly used rainfall, capital cost, and GSI performance assumptions. The results suggest that GSI investments may not be justifiable to address sewer overflow reduction alone if uncertainty is not considered.

Taking into account plausible future changes, by contrast, shows that **source reduction with GSI is more cost-effective in higher-rainfall scenarios**. Of course, all strategies show improved cost-effectiveness with more rainfall, but the relative cost-efficiency improvement for GSI-20 is high. Given that uncertainty about GSI cost and performance is also high, this suggests that **future analysis of GSI costs and benefits should take uncertainty into account** and cannot rely on point estimates alone.

The analysis also shows that source reduction investments (similar to those evaluated here) could provide long-term value, even if their investment performance compares unfavorably with that of gray infrastructure under current conditions or historical hydrology alone. Specifically, source reduction could provide near-term hedging value for ALCOSAN and municipal planners against future climate and hydrology changes. If increases in the average annual rainfall appear more likely, this could make these source reduction approaches more cost-effective to implement in the long term. By contrast, the performance of gray infrastructure designed for a specific hydrology or rainfall assumption might decline if these design assumptions are exceeded, leading to poorer performance and lower cost-effectiveness. Source reduction investments also provide flexibility in terms of the timing and sequence of project implementation, as well as the potential for incremental benefits over time, when compared with large-scale gray infrastructure.

Next Steps for Stormwater Analysis to Support Planning

Plausible future changes should inform near-term planning and design. The vulnerability analysis described here suggests that the sewer overflow challenge may have already grown in recent years and could increase further under plausible assumptions about future climate, population growth, or land-use changes. Infrastructure planning and design based only on a 2003 Typical Year could yield a system that is not resilient to these changes, meaning that overflows could still occur regularly even after these investments are implemented. Related cost estimates may also be too low—if system components need to be sized to account for additional rainfall, for example, long-term implementation costs may end up higher than currently projected. These considerations suggest a need to incorporate a range of assumptions about future rainfall, project costs, and GSI performance when planning for or designing key components of the draft WWP, identifying source reduction investments for municipalities across the ALCOSAN service area, and designing and implementing a complete green infrastructure strategy for Pittsburgh.

Source reduction should be considered for benefits beyond sewer overflow reduction. This report addresses sewer overflow, a central concern that study participants identified for regional stormwater management. The analysis shows that large-scale investments in source reduction could help reduce overflow but with a wide range of uncertainty regarding cost-effectiveness and relative strategy performance. It also suggests that source reduction may not compete well with traditional sewer infrastructure investments when considering sewer overflow reduction benefits alone, depending on key scenario assumptions.

However, participants in the workshops also identified goals, such as flood risk reduction, ecosystem services, access to green space, and economic development. Stormwater source reduction and, in particular, GSI are often encouraged to help address these goals alongside water-quality improvement, but the research team was unable to evaluate these important co-benefits within the scope of this study.

Future research could build on the RDM approach while also incorporating estimates of GSI co-benefits to provide a more complete understanding of the benefits and costs of source reduction as part of regional, integrated stormwater management plans. In turn, this could help ALCOSAN, the City of Pittsburgh, and other municipalities identify solutions that are more cost-effective and yield a broader range of benefits. The analysis framework could, in general, help to support continued progress toward regional collaboration on stormwater planning.

Source reduction could help reliably reduce overflows, but additional research is needed to fully define a long-term, adaptive strategy. This analysis suggests that investments in GSI and I&I source reduction, coupled with treatment expansion, could yield cost-effective overflow reduction if future rainfall volumes increase and certain cost and performance assumptions can be realized. In addition, it suggests that investment in regional source reduction could provide hedging value for ALCOSAN and municipal planners against future rainfall increases, potentially avoiding the need to further upgrade the gray infrastructure system. These preliminary findings should be followed by further analysis.

None of the strategies considered was able to meet overflow reduction goals in any of the simulated scenarios, however, and an important next step would be to apply a similar robustness and deep uncertainty framework with an expanded range of policy levers and strategies. For instance, additional components of the draft WWP—such as further treatment expansion, new deep-tunnel interceptors, or other new conveyance and storage not evaluated in this research scope—could be combined with I&I or GSI source reduction levers to determine the level of investment needed to reliably reduce or eliminate overflows both today and decades from now. Additional analysis using deep uncertainty planning methods could also identify specific phasing for gray and green investments as part of an adaptive plan. Our analysis is suggestive, but not yet conclusive, about what a robust and adaptive strategy might entail.

Acknowledgments

This pilot study was possible only through active and sustained engagement with many partners and participants in Allegheny County, Pennsylvania, and nationwide. First, the authors would like to thank the members of the Study Partner and Stakeholder Advisor groups for their time, insights, and patience in guiding this emerging research. Individual contributors are listed in Appendix A of this report, and we are thankful to each and every one of them for helping to support this research. We gratefully acknowledge the support from Allegheny County as a key partner throughout this effort. We also received key guidance on county stormwater planning from county government staff and John Shannon of Michael Baker Jr., Inc.

The simulation analysis and technical work described in this report were the work of many hands. Tim Prevost of the Allegheny County Sanitary Authority (ALCOSAN) provided access to ALCOSAN's stormwater simulation models and supporting data sets, along with essential technical guidance throughout the process. Jeanne Clark and Alex Sciulli with ALCOSAN also provided timely and helpful feedback, along with Colleen Hughes and Edward Kluitenberg with CDM Smith, Inc. We thank Khalid Khan, now senior project engineer at CDM Smith, for serving as our simulation modeling technical adviser. The ALCOSAN and CDM Smith team collectively provided timely feedback on an early version of this report that greatly improved the final analysis.

Matt Graham with LandBase Systems, Inc., also played an important advisory role and helped ensure that the simulation modeling results were consistent with other detailed investigations under way. We also received timely inputs from the Pittsburgh Water and Sewer Authority green infrastructure team, including James J. Stitt and Katherine Camp (Pittsburgh Water and Sewer Authority) and Tom Batroney (Mott MacDonald), and further data and guidance from John Schombert and Beth Dutton (3 Rivers Wet Weather). We also would like to thank Constantine Samaras (Carnegie Mellon University) for guiding the climate downscaling analysis conducted by co-author Lauren Cook.

We gratefully acknowledge the constructive peer reviews provided by Richard Hillestad (RAND) and Joseph Kasprzyk (University of Colorado Boulder). Their comments were insightful and timely and greatly improved this final report.

This research also benefited from the support of other RAND researchers and information systems professionals, and we would like to thank Min Gong, Neil Berg, Melissa Finucane, Adrian Salas, Paul Ng, and Katheryn Giglio for their contributions. Hilary Peterson provided critical support throughout the research effort, overseeing workshop logistics and countless revisions to the draft manuscript. Robert Lempert and Debra Knopman served as RAND co–principal investigators, overseeing all three urban climate adaptation pilot efforts, and we appreciate their leadership and constructive feedback, as well as Knopman's review of this report. Finally, we would like to thank Craig Howard, director of community and economic development at the John D. and Catherine T. MacArthur Foundation, for his guidance and vision in supporting this research effort.

CHAPTER ONE

Introduction

Water and wastewater systems are large public investments, seldom noticed by urban residents, although they represent a sizable portion of residents' local tax and public utility dollars at work. State and local governments throughout the United States spent a total of $109 billion in 2014 alone on capital, operations, and maintenance of water and wastewater treatment systems (Congressional Budget Office, 2015, p. 8), and the bill for modernizing these systems is growing. Top of the list for modernization are the combined sewer systems.

In many of the oldest cities in the United States, including Pittsburgh, combined sewer and stormwater systems were first constructed in the early 19th century to remove wastewater and rainfall from city streets (3 Rivers Wet Weather, undated-f; Burian et al., 2000, p. 43). Combined sewer systems use the same network of pipes to carry wastewater, rainwater, and snow melt away from cities. They were developed in an era before modern wastewater treatment, when stormwater was used to dilute sewage flows and reduce the frequency of waterborne disease. What seemed like a major sanitation and engineering advancement at the time, however, came to be recognized as damaging to the quality of streams and rivers and their downstream communities. In the early 20th century, U.S. cities began building separate sewer and stormwater systems, but not until the passage of the 1972 Clean Water Act (CWA) did the federal government have the legal authority to require local governments to reduce the water pollution from these combined sewer overflows (CSOs). Ever since, many urban areas throughout the east and midwest of the United States have been wrestling with the engineering and legal challenges posed by these legacy systems. However, the problem defies easy solutions when consensus is required across jurisdictions with different equities, resources, and interests in play.

This pilot study is intended to make a contribution toward addressing this challenge in Pennsylvania's Allegheny County, with a particular focus on how to address increasing vulnerabilities to high levels of overflow exacerbated by long-term uncertainties posed by a changing climate and other relevant conditions affecting the region's stormwater flows. The study builds on recent RAND efforts to support improved decisionmaking in response to climate and other uncertainties (e.g., Fischbach et al., 2015) and is one of three pilot studies centered around a new framework for urban adaptation and response to climate change (Knopman and Lempert, 2016). The goal is to support improved stormwater, wastewater, and climate resilience planning at the county, city, and municipal levels.

Allegheny County's Stormwater Challenge

Allegheny County, the center of the Pittsburgh metropolitan region in the commonwealth of Pennsylvania, is home to more than 1.2 million residents living in 130 independent municipalities, the most of any county in the state (U.S. Census Bureau, 2015). Between its famous "three rivers" (the Allegheny, Monongahela, and Ohio)[1]—the primary reason for Pittsburgh's initial settlement and its industrial-era economic dominance—and countless creeks and streams draining its steep and hilly topography, the county is dominated by water. The region also receives nearly 40 inches of rain on average per year (National Research Council, 2005).

Presently, the county and its largest municipality, the City of Pittsburgh, face a major challenge in effectively managing this natural resource through infrastructure and policy practices. Much of the water-management infrastructure—in particular, its stormwater and wastewater management infrastructure—requires substantial upgrades and reinvestments to meet U.S. Environmental Protection Agency (USEPA) water-quality standards under the CWA. The county's stormwater and wastewater infrastructure consists of multiple interconnected sewer systems. Flows from 83 of the 130 municipalities in the county, including Pittsburgh, drain into a network of pipes and tunnels leading to a single treatment plant operated by the Allegheny County Sanitary Authority (ALCOSAN) (ALCOSAN, 2012g; Tarr and Yosie, 2003). Figure 1.1 shows the extent of ALCOSAN's service area in Allegheny County, which also serves as the study boundaries for this report.

The ALCOSAN system was designed for another era of sanitation, when diluting wastewater before draining into streams and rivers was state of the art. Now, the system is an aging patchwork across many municipalities, including poorly maintained and leaking or broken pipes.

In some cases, streams that previously flowed naturally through the region's valleys and waterways are now buried and flow directly into the system and to the wastewater treatment plant (WWTP). In addition, the ALCOSAN system is inadequately sized to capture and treat most "wet weather" events (rain and snowfall), which occur frequently throughout the year. As a result, nearly every time it rains, a sewer overflow occurs in at least one of the approximately 450 outfalls somewhere in the system, draining a mix of untreated stormwater and wastewater into one of the three rivers. Some portions of the county have sanitary sewer overflows (SSOs), where wastewater flows

---

[1]   In addition, there is a fourth river, the Youghiogheny River, which is a tributary to the Monongahela.

Figure 1.1
Study Region and Extent of the ALCOSAN System

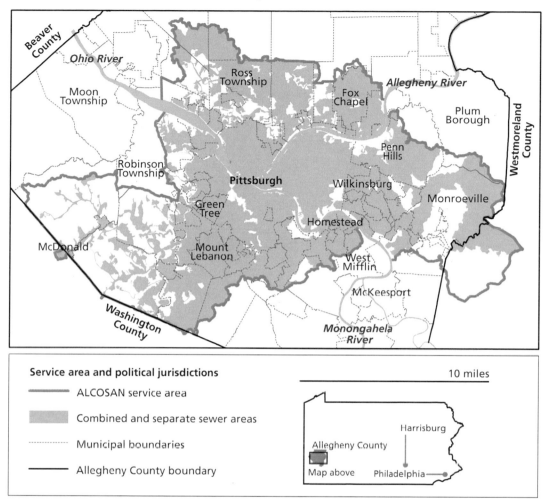

NOTE: Data from Allegheny County Sanitary Authority.

RAND RR1673-1.1

directly into streams and rivers because of the infiltration of elevated groundwater or unintended stormwater inflow. CSOs in other areas result in a mixture of wastewater and stormwater entering waterways. Total CSO and SSO overflows are estimated to exceed 9 billion gallons (Bgal.) in a typical year, representing one of the most significant overflow problems in the United States (ALCOSAN, 2012g).

Given the negative effects on water quality, ecosystems, and public health from these overflows, regulators have sought for decades to reduce or eliminate them. After a long period of negotiation, USEPA and its state enforcement partner, the Pennsylvania Department of Environmental Protection (PADEP), working with the Allegheny County Health Department (ACHD), issued a consent decree (CD) in 2008 requiring

ALCOSAN to meet water-quality requirements mandated by the CWA and state and county laws (ACHD, 1997; Comebemale et al., 2016; USEPA, 2007). Responsibility for compliance with the CD rests with ALCOSAN. However, because ALCOSAN owns or directly manages only 92 of the approximately 4,000 total miles of pipes in the system (ALCOSAN, 2012b, p. 3-5), the responsibility to manage and improve the overall system is shared by ALCOSAN and the 83 contributing municipalities, including Pittsburgh. To meet this legal mandate, ALCOSAN, in coordination with the municipalities, is required to make substantial improvements to the system by 2026 to eliminate SSOs and greatly reduce the frequency and volume of CSOs during wet weather events.

In addition to the water-quality challenges, large precipitation events and the county's hilly topography can lead to flooding in many low-lying areas. In August 2011, for example, four people were killed in a flash flood when a major thoroughfare rapidly filled with 9 feet of floodwater during an intense rainfall event caused by, in part, inadequate stormwater drainage (Balingit, 2013). Some municipalities regularly face rainfall flooding, which can damage homes, businesses, and municipal infrastructure and can increase the risk of landslides.

The Pittsburgh region's stormwater management and infrastructure challenge could grow with a changing climate or other long-term trends. Recent projections suggest that the northeastern United States is expected to see increased precipitation in a warming climate, which could manifest as an increase in the average annual rainfall, a higher number of severe rainfall events, or both (Melillo, Richmond, and Yohe, 2014; Shortle et al., 2015). A changing climate challenges the practice of using historical observations alone as a basis for future water infrastructure planning and design (Committee on Adaptation to a Changing Climate, 2015; Mamo, 2015; Milly et al., 2008). The planning challenge could be further exacerbated by population growth and land-use changes, leading to increased wastewater flows from new customers or a continued conversion of forest to impervious cover in areas with new development, which would tend to increase stormwater runoff.

As a result, ALCOSAN, Pittsburgh, and neighboring municipalities do not just need to address today's water management problems—they need to identify and plan for emerging challenges so that the system will be resilient to possible changes and future conditions.

The Long-Term Planning Challenge Is Unresolved

ALCOSAN responded to the USEPA CD by developing a draft Wet Weather Plan (WWP) in 2012. In this planning effort, ALCOSAN formulated several variations of a "gray" infrastructure plan—one that relies on traditional engineered wastewater infrastructure, such as treatment, pipes, tunnels, and storage tanks—to address

USEPA requirements of eliminating SSOs and substantially reducing CSOs by 2026. The major components of the ALCOSAN plans generally included an expansion of the treatment plant capacity and construction of new conveyance tunnels along the three rivers, designed to store substantial volumes of storm and wastewater.

The full-scale Selected Plan would meet USEPA requirements but, at a price of $3.6 billion, was deemed unaffordable for ratepayers. ALCOSAN instead proposed its "Balanced Priorities" plan, with an estimated cost of $2 billion, which would be a first step toward addressing regional sewer overflows. However, the regulators did not accept the plan because it did not control sufficient levels of overflow, and they requested further modifications in 2014. USEPA acknowledged the unaffordability of the $3.6 billion plan for ratepayers, but it pressed for further discussions to identify a plan that would be in full compliance with the CD.[2] As of this report's writing, negotiations continue among the key parties.

One key point of negotiation is the proposed inclusion of "source reduction"—that is, investments to reduce or prevent groundwater or stormwater from entering the existing sewer system "at the source," thereby freeing up capacity and reducing overflows. Source reduction approaches proposed for ALCOSAN communities include repairs to existing sewer pipes to reduce the flow of stormwater or groundwater into leaky pipes, underground stream removal from the sewer system, and such "green" approaches as green stormwater infrastructure (GSI). GSI is intended to divert and capture water during rainfall events using vegetation, soils, or other natural elements— rather than pipe stormwater into tunnels and then pump and treat it—with the goal of (1) eventually infiltrating into the ground rather than flowing as runoff; or (2) temporarily storing rainwater and slowly releasing it back into the system, avoiding or reducing the severity of a potential CSO or flood event (ALCOSAN, 2015b; USEPA, undated).

The City of Pittsburgh and many local stakeholders support the broader use of GSI to augment or offset portions of the WWP. ALCOSAN has responded with a detailed study of potential stormwater source reduction options, along with investment in a range of pilot efforts (ALCOSAN, 2015b). Pittsburgh Mayor William Peduto and Allegheny County Executive (ACE) Rich Fitzgerald have also requested permission to pursue a more-flexible, adaptive approach that utilizes GSI, with an extended window to explore stormwater reduction options (Office of the Mayor William Peduto, 2016). Relatedly, the Pittsburgh Water and Sewer Authority (PWSA) has explored GSI options and has developed its own GSI Master Plan focused on 30 key sewersheds[3] within the city boundaries (PWSA, 2016). However, these regional planners lack key information regarding how stormwater source reduction options could perform at scale or how they compare with the performance of traditional gray wastewater infrastructure.

---

[2]   Telephone interview with Jeanne Clark of ALCOSAN, September 1, 2016.

[3]   A *sewershed* is an area of land in which all wastewater flows drain to a single point. This is analogous to the drainage of surface water runoff in a watershed.

Study Purpose and Scope

Decisionmakers and planners in Allegheny County, including the ACE's office, ALCOSAN, the Pittsburgh mayor's office, PWSA, and other municipal leaders, currently do not have access to critical information to support stormwater decisionmaking at the county, city, or municipality level. Information about future changes and vulnerability is lacking, as is a framework in which to evaluate a range of proposed stormwater management options and identify cost-effective solutions that could be more robust to future conditions.

Key guiding questions for this analysis include the following:

- How might the region's vulnerability to future stormwater runoff and sewer overflow change with a changing climate and population patterns?
- To what extent could CSOs and SSOs be reduced using innovative approaches, either in current conditions or with future changes?
- How do GSI and other source reduction solutions compare with traditional "gray" solutions in terms of overflow reduction benefits, costs, and cost-effectiveness?
- What trade-offs must be resolved to implement improved stormwater management across the county?

The analysis described in this report is intended to provide a baseline of scientific information to support ongoing regional coordination around stormwater source reduction and wet weather planning. The central goal is to consider long-term and systemwide changes that might emerge from external drivers in the future, as well as the potential benefits and costs of implementing source reduction strategies at scale.

Our report represents an early step toward integrated watershed management and planning across the Allegheny County region. It is intended as an exemplar to show how estimating regionwide benefits and costs can help infrastructure and municipal planners make better near-term decisions and identify the appropriate incentives to support a regional, coordinated approach. These methods and insights could be applied in other places in the United States.

Organization of This Report

This report is organized around six chapters. Chapter Two provides additional background on the planning context and regional sewer system, describes the overall approach used in this investigation, and summarizes the outcomes from initial planner and stakeholder workshops. Chapter Three details plausible future vulnerabilities that could emerge in the sewer system with future change if no additional actions are taken. In Chapter Four, preliminary simulation modeling results from a range of source reduction policy levers are presented, with the goal of informing a more-discrete set of strategies to consider in detail. In Chapter Five, the benefits, costs, and cost-effectiveness of a subset of source reduction strategies are evaluated using a range of scenarios representing current and future uncertainty about wastewater customer demand, climate and land-use changes, and infrastructure capital costs. Finally, Chapter Six concludes with a brief summary of findings and suggested next steps for research and planning.

In Chapters Three through Five, the approach and methods are described briefly, with additional detail reserved for a series of technical appendixes. The study is supported by five appendixes, available online (www.rand.org/t/RR1673). Appendix A provides a full list of participants in study workshops and the deliberative process. Appendix B describes how we adapted a series of stormwater and wastewater simulation models originally developed by ALCOSAN to support quantitative scenario analysis. Appendix C details the technical inputs and methods applied to develop scenarios for this study. In Appendix D, we describe the policy levers and planning-level strategies tested in the analysis, including improvements to the existing sewer system and different source reduction options. Finally, Appendix E details the final experimental design utilized in Chapter Six for the Robust Decision Making (RDM) analysis.

CHAPTER TWO

Regional Stormwater Planning in Allegheny County

This report opened with a brief description of how aging and undersized stormwater and wastewater infrastructure, coupled with the potential for future uncertainty, threatens water quality, recreation, and other benefits from Allegheny County's extensive river and stream network. This chapter gives deeper context to the problem faced by ALCOSAN and municipal planners responsible for stormwater management. We present the regional stormwater challenge as identified and discussed by study participants through a series of workshops and then describe how these discussions were translated into technical inputs to support the analysis. The chapter also describes the participatory framework and methods for decisionmaking under deep uncertainty used to identify and evaluate infrastructure management strategies across different futures to understand which combinations could make stormwater and wet weather planning in Allegheny County more robust.

Water Resources Are Vital to Allegheny County

Allegheny County and the Pittsburgh metropolitan region are shaped and defined by water. The Allegheny and Monongahela rivers (and their tributaries) converge in Pittsburgh to form the Ohio River, and the entire county is within the Ohio River basin. Major state transportation routes generally follow the paths of the rivers, and navigation along the rivers to support commerce has been a defining feature of Pittsburgh's economy since the city's founding. Most of the public water supply is also drawn directly from these surface waters.

The three rivers are also used daily by residents for recreation activities, such as fishing, boating, kayaking, biking, and swimming. Waterways are intrinsically linked to the region's culture, serving as a backdrop to Pittsburgh's sports stadiums and shuttling riverboats full of visitors taking historical tours of the area. The rivers support Allegheny County's economy through one of the busiest inland ports in the nation and have been identified as an important part of the area's postindustrial redevelopment (Collins et al., 2005; National Research Council, 2005; River Life Task Force, 2001).

More than 2,000 miles of streams and 90 miles of rivers flow through Allegheny County. Figure 2.1 shows a map of the 25 watersheds within the county's border. However, urbanization and the region's long history with natural resource extraction and heavy industry have taken a major toll on the health and function of the county's waterways. PADEP classifies nearly half of these waterways—more than 940 miles—as impaired, with mine drainage and stormwater runoff representing the most frequent known contributors to stream impairment. Stormwater is the second-largest single contributor to stream impairment, affecting nearly 20 percent of impaired water bodies (Michael Baker Jr., Inc., 2014, pp. 11–12).

Figure 2.1
Watersheds in Allegheny County

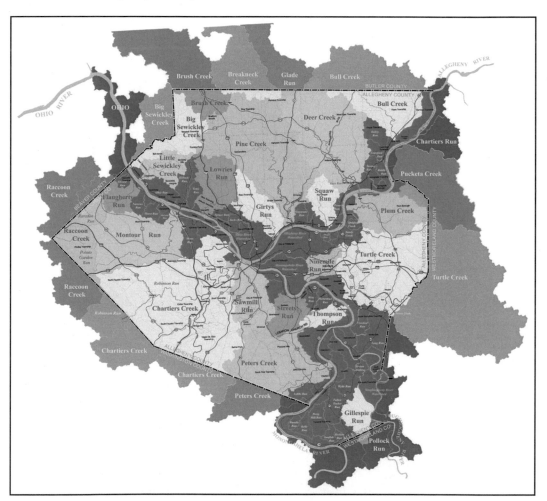

SOURCE: Allegheny County Conservation District.

There is no single overall authority responsible for the management of stormwater issues or water quality in Allegheny County. Some of this is rooted in the Commonwealth of Pennsylvania's and Allegheny County's historical experience with "home rule," in which authority on a range of governance issues is devolved from the state to the county or local level (Allegheny County, 2000; Institute for Public Policy and Economic Development, 2009). It also derives from the metropolitan region's long and complex history of incremental progress toward wastewater treatment and water-quality management, described in detail in other sources (National Research Council, 2005; Tarr and Yosie, 2003; Yosie, 1981).

Overview of the ALCOSAN Sewer System

Eighty-three municipalities, including the City of Pittsburgh, operate within the ALCOSAN service area, which makes up roughly two-thirds of Allegheny County (see Figure 2.2) (ALCOSAN, 2012g).[1] The outer portions of the service area were generally constructed more recently and consist of *separated sewer systems*—that is, there are separate pipe networks for stormwater and sewage. In a separated sewer area, stormwater is generally routed directly to a nearby river or stream, while sewage from houses and other buildings is routed through the sanitary sewer system to the ALCOSAN treatment plant. Separated sewer systems are the preferred modern design. However, in many of the older parts of the region—primarily within Pittsburgh city limits—there is a *combined sewer system*, in which both stormwater and wastewater flow through the same pipe network.

Parts of the combined sewer system are more than 100 years old, relics from the early 20th century, when untreated sewage was simply routed to the waterways. ALCOSAN was formed in 1946; and, in the 1950s, it constructed a single treatment plant and installed approximately 92 miles of large interceptor sewer pipes that are still in use today. Locations where combined sewers previously drained into rivers were connected to these interceptor pipes, which provide conveyance to the treatment plant. However, when the capacity of the system is overwhelmed, typically during wet weather events, a mixture of untreated stormwater and sewage overflows from the system and is released into a river or stream. Importantly, many of the outlying regions with separated sewer systems flow into the combined sewers to reach the ALCOSAN treatment plant (3 Rivers Wet Weather, undated-e). As a result, overflows in the combined area may also include wastewater from the upstream separated areas.

In total, the sewer system serves a population of roughly 900,000 and consists of more than 4,000 miles of municipally owned and operated pipes, in addition to the

---

[1] Allegheny County municipalities outside of the ALCOSAN service area contribute to other smaller storm- and wastewater collection systems or have stand-alone septic systems.

Figure 2.2
ALCOSAN System Map

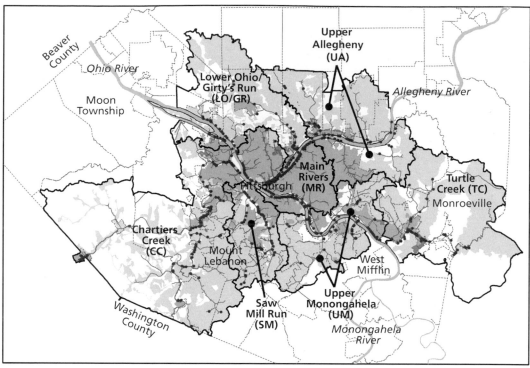

SOURCE: Data from Allegheny County Sanitary Authority.
NOTE: SWMM = Storm Water Management Model.

major trunk lines and interceptors owned and operated by ALCOSAN. There are 448 outfalls throughout the system, each a potential location for CSOs or SSOs. The sole treatment plant for the ALCOSAN service area is Woods Run, which is on the Ohio River roughly four miles northwest from downtown Pittsburgh. The current plant capacity is 250 million gallons per day (MGD), translating to roughly 90 Bgal./year. The plant is operating near full capacity on a daily basis, treating approximately 190 MGD (ALCOSAN, 2012b, p. 3-7).

Rainfall volumes of as low as 0.1 inches can overwhelm the capacity of the system, resulting in overflows (3 Rivers Wet Weather, undated-b). Past analysis by ALCO-

SAN has found that there are roughly 9 Bgal./year in overflows from the CSOs and 600 million gallons (Mgal.) per year in overflows from the sanitary sewers in the separated areas (SSOs) (ALCOSAN, 2012g). Although the volumes are much lower, SSOs can be a more serious concern because there is a higher concentration of wastewater; with CSOs, the wastewater is generally heavily diluted with stormwater. Total overflow in the ALCOSAN system are more than 10 percent of the annual volume treated at the Woods Run treatment plant.

Overflows in Pittsburgh and other ALCOSAN communities are high, but they are comparable to those in older American cities with combined sewer systems. Milwaukee, Wisconsin, has a similar-sized system, for example, and previously faced 8 to 9 Bgal./year of CSOs. After spending approximately $4 billion on system improvements since the 1980s, including the construction of a deep-tunnel interceptor and a series of stormwater management projects, overflows have been reduced to 1.4 Bgal./year (City of Milwaukee, undated). New York City, which has a massive system with 14 treatment plants serving millions of customers, faces annual overflows of roughly 30 Bgal./year (New York City Department of Environmental Protection, undated). Missouri's Kansas City, which has a wastewater system serving more than 650,000 customers, faces an average annual overflow volume of 6.4 Bgal./year. Its overflow control plan, which aims to capture 88 percent of the these flows, is estimated to cost $2.5 billion by 2035, constituting the largest infrastructure project in the city's history (Kansas City, Missouri, Water Services Department, 2012).

Regional Stormwater and Sewer System Planning Efforts

In recent decades, there have been a range of proposals and planning efforts intended to address the fragmentation challenge and guide municipalities in Allegheny County toward a regional approach. For example, ALCOSAN and the Allegheny Conference on Community Development convened a Sewer Regionalization Review Panel in 2011, chaired by former Carnegie Mellon University president (now president emeritus) Jared Cohon. Finalized in 2013, the panel's findings called for a range of actions to improve collaboration and integration, including ALCOSAN governance changes, financial incentives for municipal flow reduction, voluntary system consolidation, and integrated planning (Allegheny Conference on Community Development, 2013). Carrying one set of recommendations forward, 3 Rivers Wet Weather subsequently partnered with the Congress of Neighboring Communities (CONNECT) to assemble a Sewer Regionalization Implementation Committee (SRIC) made up of county and municipal planners, ALCOSAN staff, stormwater experts, and stakeholders. In 2015, the committee produced a final report that notably detailed a process for the transfer of ownership and responsibility for approximately 200 miles of intermunicipal trunk sewers and selected other facilities from individual municipalities to ALCOSAN (Sewer

Regionalization Implementation Committee, 2015). 3 Rivers Wet Weather proposed that municipalities develop strategies that cross boundaries and consider the issue on a regional, rather than municipal, scale to lessen the burden that the multibillion-dollar public works project would have on the residential ratepayers (3 Rivers Wet Weather, undated-a).

ALCOSAN has similarly initiated meetings involving policymakers, managers, and engineers to facilitate municipal coordination on green infrastructure projects. These meetings served to highlight the varying desired outcomes among municipalities while generating discussion on possible funding structures and opportunities for cooperation (ALCOSAN, 2015b, p. 4-17; Allegheny County, 2000). ALCOSAN provided customer municipalities with access to professional consulting firms and technical support to aid in stormwater planning, in addition to meeting facilitation. Building on and supporting these collaborative efforts are likely to be key steps in the development of a regional strategy.

Next, we summarize several key planning efforts conducted by government agencies that are under way and speak directly to the stormwater and wastewater management challenge.

ALCOSAN Wet Weather Plan

As a result of aging wastewater infrastructure, recent decades have seen an increase in both sewer overflows and flooding issues related to system capacity limits. On January 23, 2008, ALCOSAN entered into a CD with USEPA, PADEP, and ACHD, which required that a sewer overflow long-term control plan (LTCP) be drafted and submitted to the agencies by January 30, 2013. The CD required ALCOSAN to create an implementation plan for the elimination of SSOs and the reduction of CSOs to meet federal CWA requirements, consistent with the CSO policy. The final goal of the CD was to improve water quality in the region and preserve the use of waterways for drinking water and recreation. Under the original CD, the proposed infrastructure must cover the needs of the region through the year 2046, and construction of all necessary components must be completed by 2026 (ALCOSAN, 2012g, p. 1-1).

The resulting draft WWP, designed to meet the goals of eliminating SSOs and mitigating CSOs by the 2026 deadline, consists of a range of large-scale gray infrastructure components, including expanded treatment capacity, new conveyance sewers, and deep tunnels along each of the three rivers intended to store additional combined sewer flows during rainstorms. The Selected Plan, which mainly consists of these gray infrastructure investments, is anticipated to reduce CSO volumes by 92 percent while completely eliminating SSO volumes (ALCOSAN, 2012d). The resulting effect would avert approximately 9 Bgal. of wastewater per year from reaching waterways.

Concerned about how the ratepayer base could absorb the $3.6-billion price tag for the Selected Plan, ALCOSAN then recommended a less-costly, near-term compromise that focuses on eliminating most SSO outflows while reducing the number and volume of CSOs. The lower-cost alternative ($2 billion), the Balanced Priorities plan, is projected to prevent more than 5 Bgal. of wastewater from reaching waterways each year, yielding systemwide reductions in CSO and SSO volumes of 56 percent and 90 percent, respectively. ALCOSAN estimated that ratepayers could face rate increases of approximately 10 percent per year to fund this plan (ALCOSAN, 2012g).

The Balanced Priorities plan, to which we refer as the "draft WWP" through the remainder of this report, focuses on seven key elements, including the expansion of treatment plant capacity to 480 MGD; construction of storage and conveyance tunnels extending from the treatment plant along the Ohio, Allegheny, and Mononga-hela rivers; and the implementation of hydraulic improvements throughout the sewer system (ALCOSAN, 2012e).

ALCOSAN submitted the draft WWP to USEPA in early 2013 and simulta-neously requested an 18-month extension to further investigate the value that GSI could provide to the region. Public stakeholders also raised concerns that the pro-posed strategy did not fully explore alternatives, such as green infrastructure.[2] Some stream removal and restoration projects have already been completed by ALCOSAN, and, in the near future, ALCOSAN plans to support investment in GSI by customer municipalities through a funding program (ALCOSAN, undated). In January 2014, USEPA informed ALCOSAN that the draft WWP would not address all water-quality improvements outlined in the 2008 CD and requested further modification (Hopey, 2014).

In response to requests from the City of Pittsburgh and Allegheny County, USEPA suggested in March 2016 that it would consider extending the currently agreed-upon 2026 implementation deadline, with the goal of reducing the weight of the expendi-ture on ALCOSAN customers and moving toward a phased or adaptive approach to the overflow problem. As part of the extension, USEPA proposed that an expansion of the ALCOSAN treatment plant would be the first priority, followed by the implemen-tation of GSI, stream removal, or other investments within customer municipalities to reduce flows "at the source" (i.e., source reduction). The response from USEPA pro-poses a deadline of September 2032 for the construction of municipal flow reduction projects, as well as the ALCOSAN WWP (Hopey, 2016). As of this report writing, negotiations between the regulators, ALCOSAN, and the municipalities are still in progress (ALCOSAN, 2012g, p. 14).

Municipal Planning Efforts

Separately from the CD issued to ALCOSAN, municipalities in the ALCOSAN ser-vice area were also evaluated by USEPA for compliance with the CWA. Municipalities with SSO violations received administrative consent orders from the ACHD, while those with CSO National Pollutant Discharge Elimination System permits entered

---

[2]   Telephone communication with Jeanne Clark of ALCOSAN, September 1, 2016.

consent orders and agreements with PADEP. Municipalities facing CWA compliance violations participated with ALCOSAN in the creation of the WWP to help ensure that it would address all wastewater issues facing the service area (ALCOSAN, 2012g, p. 1-8).

To develop a systemwide WWP, ALCOSAN partitioned the service area into seven planning basins (ALCOSAN, 2012g, p. 1-9). Some areas of the municipal sewer systems are more than a century old, while other sections have been recently installed, creating a patchwork network with varying capability. Data gathered from field investigations conducted by municipalities were cataloged to digitize the physical sewer system. These data sets were then used to create a simulation model for each planning basin that characterized system behavior and overflows resulting from precipitation and customer inflows (ALCOSAN, 2012b, p. 3-12). The process of simulation model development is further described in Appendix B of this report (ALCOSAN, 2012g).

In March 2015, PADEP granted all ALCOSAN customer municipalities new 18-month consent orders and agreements. PADEP requested that municipalities use the extension to conduct a source reduction study including GSI or pipe inflow and infiltration (I&I) flow reduction. The new consent orders and agreements require a demonstration project to construct and monitor a GSI or I&I pilot project within the jurisdiction of each municipality or jointly with multiple municipalities within a common sewershed. The recently issued consent orders and agreements allow municipalities to continue issuing sewer taps for new or redevelopment construction on the condition that the municipality is in compliance with its consent order (PADEP, 2015).

Allegheny County Stormwater Management Planning

Pennsylvania General Assembly Act 167—known as the Stormwater Management Act of October 4, 1978—calls for stormwater management in the state to be conducted at the watershed level. Under Act 167, every county is required to develop stormwater management plans (SMPs) for its watersheds (Commonwealth of Pennsylvania, 2008). An SMP for each individual watershed is developed through coordination of multiple municipalities working together through a Watershed Plan Advisory Committee (WPAC). Act 167 aims to align planning with responsible land-use and water resource management and serves as the vehicle by which management programs are authorized to develop comprehensive plans addressing flood management, restoration of historical runoff pathways, and conservation of groundwater resources (Allegheny Places, undated, p. 2).

Allegheny County recently completed Phase I of its comprehensive SMP, an independent process separate from ALCOSAN's WWP development. Phase I sets out the goals of the overall project, provides a survey of watershed characteristics, identifies significant problems, and considers potential solutions for follow-up in Phase II. The final plan will focus on developing a model ordinance for stormwater management and other key elements highlighted in Phase I, but it could also contribute to subsequent watershed implementation plans and use a structure that allows for integration with a larger watershed plan in the future (Michael Baker Jr., Inc., 2014).

Pittsburgh Green Infrastructure Plan

The City of Pittsburgh, led by PWSA, has evaluated GSI options to develop its own GSI Master Plan. The plan identifies 30 priority watersheds within the city and provides recommendations for implementing GSI to achieve source reduction. The draft citywide plan was made available in fall 2016 for public comment (PWSA, 2016).

The process to develop Pittsburgh's plan focused on 30 "high-priority" sewersheds within the city and involved evaluating a combination of pilot implementation sites, specific implementation sites under consideration, and conceptual locations for GSI.[3] The goal of the assessment was to identify a GSI-based integrated planning approach to reduce CSOs and SSOs, as well as remove or detain stream inflows, mitigate specific flood hazards, and reduce the occurrence of basement sewage backups. The analysis developed a stormwater overlay lens for use as a comprehensive planning tool for new development and redevelopment opportunities in the future.

The plan is intended to address untreated overflow volume reductions while also identifying GSI co-benefits, such as community amenities and water-quality improvements (PWSA, 2016). In advance of this plan, PWSA published a document entitled "Greening the Pittsburgh Wet Weather Plan" in 2013 (PWSA, 2013). The document was informed by a series of charrettes with local and national stormwater management experts and provides recommendations to guide planning and development of GSI projects to address challenges throughout the service area. In addition to planning recommendations, the process resulted in 17 grants awarded in 2015 to local pilot GSI projects.

The draft PWSA City-Wide Green First Plan helps to build regional knowledge regarding the potential benefits of large-scale GSI implementation and source reduction. The report indicates that the most cost-effective first step for the city is the expansion of the current treatment plant, along with the maintenance of existing deep tunnels to increase the system conveyance. Following those upgrades, PWSA suggests the incorporation of large-scale GSI based on performance of initial smaller local projects. PWSA concluded that informed implementation of GSI could reduce overflows, curb flooding risks and basement backups in specific areas, and provide additional environmental and social benefits (PWSA, 2016).

---

[3]   Personal communication with James Stitt of PWSA, August 16, 2016.

Stormwater Source Reduction and Green Stormwater Infrastructure

As noted in the previous section, one approach under active consideration for ALCOSAN communities is to use GSI as an alternative management approach to traditional gray engineered pipe conveyance networks. GSI can include permeable pavement, rain gardens or rain barrels, green roofs, street planters, vegetative swales, and bioretention or infiltration trenches (Figure 2.3). These necessarily operate at multiple scales, ranging from the individual homeowner or institutional building to large-scale, centralized projects on publicly owned spaces. GSI projects are intended to divert and capture water during rainfall events, either temporarily (with eventual flow back to the system) or permanently (through infiltration or evapotranspiration) using vegetation, soils, or other natural elements (USEPA, undated). By delaying or eliminating runoff that would otherwise enter the combined sewer system, GSI can potentially reduce overflows. Some types of green infrastructure can also help improve water quality by naturally filtering stormwater runoff, which may include nitrogen from fertilizers or pollutants and sediment from roads and parking lots.

Figure 2.3
Examples of Green Stormwater Infrastructure

| Permeable pavement | Rain barrel | Green roof |

| Vegetative swale | Bioswale | Infiltration trench |

SOURCES: For permeable pavement, Creative Commons photo by Center for Watershed Protection via New York State Stormwater Flickr, 2015; for rain barrel, Creative Commons photo by Mark the Trigeek, 2011; for green roof, Creative Commons photo by David L. Lawrence Convention Center via Flickr (used with permission), 2013; for vegetative swale, Creative Commons photo by Louis Cook, Philadelphia Water Department, via Flickr, 2013; for bioswale, Creative Commons photo by Aaron Volkening via Flickr, 2010; for infiltration trench, Creative Commons photo by Philadelphia Water Department via Flickr (used with permission), 2012.
RAND RR1673-2.3

Advocates for GSI also point to a range of other potential co-benefits to local communities, including additional green space and community amenities, increased property values, improved air quality, reduced urban heat island effects, or community development and economic growth opportunities (Gaffin, Rosenzweig, and Kong, 2012; Gill et al., 2007; USEPA, 2014). However, because GSI is a relatively new approach for controlling sewer overflows, there are limited examples of large-scale GSI implementation with proven results. There are also concerns that GSI will lose its effectiveness over time if not properly maintained.

USEPA supports the use of GSI as a component of future long-term sewer overflow control plans (USEPA, 2014), and the approach is being studied and implemented at scale as part of LTCPs or other climate adaptation efforts in other major metropolitan areas, including Philadelphia; Seattle; Washington, D.C.; and New York City (District of Columbia Water and Sewer Authority, 2015; New York City Department of Environmental Protection, 2010; Philadelphia Water Department, 2009).

Deliberation with Analysis to Support Stormwater and Wastewater Planning

Given the high number of municipalities, decisionmakers, and stakeholders involved in county stormwater management, a major component of this study was a deliberative process with a representative group of regional decisionmakers and stakeholders. Working within the framework described in Knopman and Lempert (2016) the process involves facilitated discussions about regional goals and acceptable metrics for measuring progress toward goals, development of options for addressing the problem or problems at hand, integration of existing data and models to represent the physical and socioeconomic systems of interest, and a facilitated discussion of alternative strategies and trade-offs across goals.

Building on the related stormwater and land-use planning efforts under way or recently completed, we worked in close collaboration with Allegheny County, ALCOSAN, the City of Pittsburgh, PWSA, and other local partners to conduct a participatory planning exercise and address the guiding questions described in Chapter One.

Methodology

Our approach used a "deliberation with analysis" process of decision support, in which parties to the decision deliberate on their objectives, options, and problem framing; analysts generate decision-relevant information using the system models; and the parties to the decision revisit their objectives, options, and problem framing influenced by this quantitative information (National Research Council, 2009).

To this end, we employed RDM, a quantitative decision analytic framework for supporting decisionmaking under conditions of deep uncertainty. Deep uncertainty occurs when experts, decisionmakers, and stakeholders do not know, or cannot agree on, the likelihood of future conditions or the relative importance of key drivers or outcomes from a given decision (Groves and Lempert, 2007; Lempert, Popper, and Bankes, 2003; Lempert and Collins, 2007).

Rather than using models and data to describe a best-estimate future, RDM addresses deep uncertainty by running the analysis backward. Typically, this entails using "exploratory modeling," or the systematic employment of one or more simulation models over hundreds to thousands of different sets of assumptions to describe how plans perform in many plausible futures. The approach is intended to leverage computer simulation to systematically consider how outcomes could vary across a wide range of plausible assumptions about the future.

RDM then uses statistics and visualizations to investigate the resulting large database of model runs and help identify those model assumptions and future conditions in which plans will perform well or poorly (Bankes, 1993; Bryant and Lempert, 2010; Weaver et al., 2013). This portion of the investigation, known as scenario discovery, seeks to identify a small set of key uncertain factors or a subset of the range of uncertainty that would lead key parties to a different decision if they were to occur. In other words, scenario discovery helps translate an ensemble of quantitative simulations into a manageable number of understandable and meaningful "decision-relevant" scenarios to support deliberations.

RDM is one of a set of decision support methods that share a common core of exploratory modeling, iterative improvement in response to emerging vulnerabilities, and adaptive decisionmaking (e.g., Brown et al., 2012; Hall et al., 2012; and Walker, Haasnoot, and Kwakkel, 2013). Herman et al. (2015) provides a useful review and taxonomy of these related approaches. Relevant recent applications of RDM and related methods include water supply planning in response to climate change (Groves, Fischbach, Bloom, et al., 2013; Groves, Fischbach, Kalra, et al., 2014; Groves, Bloom, et al., 2015; Haasnoot et al., 2013; Herman et al., 2015), flood risk management (Fischbach, 2010; Woodward, Kapelan, and Gouldby, 2014), long-term coastal planning (Groves, Fischbach, Knopman, et al., 2014; Kwadijk et al., 2010), and water-quality implementation planning for several pilot watersheds (Fischbach et al., 2015). Methods for decisionmaking under deep uncertainty are described in much greater detail in this prior literature on methods and applications.

This pilot study uses the deliberation with analysis and RDM framework to consider how proposed source reduction policy levers, other proposed near-term infrastructure improvements, and combined management strategies could make wet weather planning in Allegheny County more robust to a wide range of future conditions. Our report describes a stakeholder-informed, iterative research process, conducted over an 18-month period. The result is a preliminary analysis that describes plausible future vulnerability and evaluates potential sewer overflow reduction strategies that could help to augment current plans.

Study Participants

Study partners and stakeholders took part in the interactive planning exercise, as outlined earlier. Previous analytic facilitation efforts (in, e.g., Louisiana, Colorado, and the Chesapeake Bay) conducted by RAND team members used detailed simulation models trusted by the stakeholders to relate actions (policy choices) to consequences (change in risk) and used visualization tools to help the stakeholders envision vulnerabilities and trade-offs, engage in an iterative process of "what if" with the data, and deliberate with one another over decision options.

At the outset of this process, the RAND team worked with the ACE's office to convene two groups of participants for the study process, Study Partners and Stakeholder Advisors. The Study Partners convened for five workshops during the project, from May 2015 through September 2016. We also met separately with the Stakeholder Advisors twice during this deliberation period. Participants in each workshop were asked to provide feedback through a written survey evaluation, which was conducted separately from the Allegheny County study team. Results from this evaluation will be published separately, together with evaluation results from RAND's other two pilot studies (ongoing as of this writing).

In addition to the formal workshops, we also conducted dozens of separate one-on-one and small-group meetings and presentations with study participants to elicit real-time feedback and keep key contributors informed regarding technical analysis progress. These touch points were critical to ensuring the accuracy, reliability, and relevance of the emerging technical analysis results.

The composition of each participating group is described next. Appendix A presents all Study Partners and Stakeholder Advisors who contributed to the project.

Study Partners

Study Partners consisted primarily of planners and technical experts conducting research and analysis at the county or city level. These partners directly participated in and supported the analysis and deliberation process. Partners were asked to commit staff time to attend meetings; provide data, models, or modeling support; help to identify and provide preliminary design input for source reduction and other policy levers considered in the analysis; and review written products. Study Partners also worked with us to estimate associated project-level and strategy costs.

Invited Study Partners from local government or quasi-government agencies included the ACE, which served as co-convener for the process; the City of Pittsburgh Office of the Mayor; ALCOSAN; PWSA; and the Allegheny County Conservation District.

Technical experts asked to join this group supported the analysis by providing data, modeling support, and feedback on the design inputs. These experts also provided preliminary quality assurance and review on the technical analysis. We were able to leverage their respective technical expertise, and provided analysis and recommendations as constructive feedback in turn. A number of Study Partners with technical expertise on stormwater and wastewater infrastructure management were also actively conducting or supporting local planning efforts. This group included experts from ALCOSAN; PWSA; 3 Rivers Wet Weather; Michael Baker Jr., Inc.; CDM Smith; Mott MacDonald; Landbase Systems; and Ethos Collaborative. Other technical experts came from local research universities, including Carnegie Mellon University's Civil and Environmental Engineering (CEE) Department, Penn State University's Penn State Center Pittsburgh, and the University of Pittsburgh's Institute of Politics and CONNECT.

Stakeholder Advisors

Two groups of regional stakeholders assisted with the study as Stakeholder Advisors. The first group included representatives from local municipalities and other local and regional decisionmakers with a vested interest in stormwater and wastewater planning. The second group included economic development organizations, watershed associations, environmental non-government organizations, community groups, and others with a sustained interest in the local stormwater and wet weather planning challenge. Participating Stakeholder Advisors took part in the interactive workshops, met with the RAND team separately, or both.

Deliberation with Analysis Process

Study participants contributed to the research throughout the study. The Study Partners and Stakeholder Advisors were first asked to help define the research scope and identify the inputs needed to conduct the technical analysis, following the "XLRM" approach often used at the outset of an RDM investigation (Lempert, Popper, and Bankes, 2003). The results of this scoping process are described in the next section. In parallel, we convened technical meetings with partners to identify existing modeling and data resource knowledge gaps, analytic trade-offs (e.g., geographic scope, model resolution), and a path to usable and verifiable simulation models within the study period.

In subsequent meetings and workshops, participants interacted directly with initial vulnerability and strategy performance results generated in the simulation analysis using an interactive decision support tool. In these meetings, the participants provided feedback and interpretation for the initial results, identified possible augmentations or improvements to the policy lever strategies, and discussed key trade-offs (e.g., overflow reduction and cost). The participants also identified a range of questions and next steps that we were not able to include in the pilot technical analysis but nevertheless represent an important research agenda for future integrated watershed analysis and planning in the county. We return to these next steps in the final chapter of this report.

Scoping the Analysis with the XLRM Framework

The first step in this investigation was to work with study participants and identify a joint research scope built around urban stormwater management within Allegheny County. We employed the XLRM process first set forth in Lempert, Popper, and Bankes (2003) to organize the conversations and relevant outputs. In the XLRM abbreviation, M stands for the goals to be met and associated performance metrics that are used to quantify these goals. L stands for policy levers or broader strategies that could be implemented to achieve these goals. X stands for uncertain factors that could affect the ability to achieve these goals but are outside the control of decisionmakers; these often represent deep uncertainty to which participants may assign very different likelihoods or that may be difficult or impossible to characterize with probabilistic risk analysis. Finally, R stands for the relationships among these elements as reflected in simulation or planning models, essentially serving as the means to consider how policy levers (L) and uncertain factors (X) combine to yield decision-relevant outcomes (M).

We used the XLRM framework to conduct kickoff workshops with the Study Partners and Stakeholder Advisors, respectively. The conversation from each workshop was documented in detail, and the RAND team subsequently condensed and merged the results into a single XLRM figure to help guide next steps. The initial XLRM scope is summarized in Table 2.1, and each quadrant is discussed briefly in turn in the following sections. Elements in bold in Table 2.1 represent factors we were able to incorporate directly into the technical analysis described in the remaining chapters of this report.

Goals and Metrics

Workshop participants from both groups identified a broader range of policy goals related to regional stormwater management than previously captured in recent wet weather and sewer overflow planning efforts (such as the draft WWP), including reducing flood risk, economic development, and access to community amenities and green space. This may be due to ALCOSAN not being set up as a regional stormwater authority and having a more limited mandate to provide wastewater conveyance and treatment and to meet water-quality requirements through overflow reduction.

Participants reinforced the importance of improving water quality in rivers and streams to respond to the current CD, address pressing public health concerns (i.e., SSOs), and support the continued economic and community redevelopment of the Pittsburgh metropolitan region and its many riverfronts. Related goals were infrastructure protection—ensuring that the water management system is properly recapitalized and maintained to provide reliable service and avoid future failures—and meeting potential future stormwater regulatory requirements from USEPA that could fall on individual municipalities to address.

Table 2.1
Scoping Summary from Partner and Stakeholder Workshops

Uncertain Factors (X)	Policy Levers and Strategies (L)
• **Climate change** • **Land-use, development, and environmental changes** • **Economy and population changes** • **Demand for water and wastewater services** • **Cost-effectiveness and affordability** • Regulatory and political landscape • Popular opinions and public sentiment • Stormwater and wastewater modeling uncertainty • **Infrastructure performance uncertainty**	• **Gray infrastructure (tunnels, pipes, treatment, storage)** • **Stormwater source reduction** • **Retrofitting, repair, operations, and maintenance** • **GSI** • **Regulations on land use and zoning** • Integrated ecosystem services • Centralized management organization and plan • Market-based solutions, innovative incentive design, and financing • Education, strategic communication, and public awareness

Relationships (R)	Goals and Metrics (M)
• **Hydrologic and hydraulic models** • Flood risk models • **Downscaled climate-informed hydrology** • **Land-use change model** • **Infrastructure cost estimation tools**	• **Improve water quality** • **Reduce sewer overflow** • Reduce flood risk • **Comply with regulatory requirements** • **Protect infrastructure** • Protect and improve ecosystem • Improve property values, add community amenities, and reduce risk premium • **Build regional cooperation and coordination** • Gain public support

NOTE: Responses in bold represent those that the RAND team was able to carry forward into the technical analysis.

Another key goal discussed during the workshops was reducing flood risk to residents and property, either from events that exceed system capacity (e.g., basement backups, manhole overflows) or from stormwater flow itself on hillsides and in valleys during severe rainfall events. Participants also identified improved green spaces and ecosystem protection, higher property values and community amenities, and a lower cost to residents that could emerge if overflows and flood risk are reduced.

These goals are presented in Table 2.1 as a single, consensus list, but competing perspectives and limited resources mean that trade-offs among these goals are likely. Some participants emphasized meeting regulatory requirements first, for instance, while others suggested that providing community amenities and ecosystem services should be weighted equally with regulatory and water-quality goals. That said, both discussions identified regional cooperation and coordination—which would help bring all of these goals together into a common framework and allow a range of planners and stakeholders to help resolve trade-offs in partnership—as critical to the success of future stormwater management and a worthy goal in itself. This would also entail improved resident engagement to build trust and public support.

Policy Levers and Strategies

Workshop participants identified a range of policy levers that could address these goals. These included traditional gray infrastructure (tunnels, pipes, treatment, and storage); stormwater source reduction via GSI or through improved repair, retrofitting, operations, and maintenance of the existing sewer system; and land-use or zoning regulations intended to reduce or eliminate stormwater runoff into the system from private property by mandating onsite stormwater capture, reduction in impervious area (pavement), or adoption of stormwater best management practices on new or redeveloped properties.

Other levers of note included a more holistic approach to ecosystem management—for example, consolidated maintenance of stormwater GSI as part of overall parks and green space management and centralized stormwater planning and management, moving toward an integrated watershed management approach. Participants also noted a range of emerging policy options that could help to implement these approaches, even in the current fragmented planning environment, including incentives for source reduction, market-based approaches to flow reduction (e.g., a stormwater fee), and innovative financing to better support the range of investments needed. Finally, the partners and stakeholders noted the importance of a public engagement and education campaign to undergird stormwater management strategies and build public awareness of the problem and costly proposed solutions.

Uncertain Factors

Uncertain factors that threaten the success of stormwater management plans in Allegheny County emerged from different sources. Workshop participants identified climate change as a key driver, particularly potential climate effects on future rainfall patterns across the region. They also noted that changing patterns of land use, development or redevelopment, and environmental management could also substantially contribute to the success or failure of water-quality or stormwater plans and investments. Relatedly, participants mentioned future demand for water and wastewater services, including safe and clean drinking water and potential new sewer connections in growing areas, as important drivers to consider. Land use and demand for water and wastewater will be driven, in turn, by broader trends in regional population and economic growth over the next 25 to 50 years, largely out of the control of local and regional decisionmakers.

The economic backdrop of the region also governs which approaches may or may not be affordable for sewer ratepayers in the region, a key concern in the current CD negotiations. Workshop participants also noted the importance of the regulatory and political backdrop at the regional, state, and national levels; for instance, new federal funding made available for infrastructure investment could dramatically alter the range of feasible solutions for local and regional planners.

Finally, some participants noted that some proposed stormwater policy levers, particularly stormwater GSI, have not yet been tested or proven at a large scale and thus have widely uncertain benefits and costs. These levers are also challenging to simulate accurately and reliably in existing quantitative tools and models, adding a further layer of uncertainty when trying to assess performance or compare with other infrastructure approaches. We return to GSI performance uncertainty in Chapter Five of this report.

Relationships

Participants only touched briefly on the system-level relationships between the factors noted earlier, largely in response to RAND team questions regarding simulation model availability to support this pilot investigation. That said, the types of models needed for regional stormwater planning noted by participants included hydrologic and hydraulic (H&H) models, stormwater runoff and flooding analysis tools, downscaled climate projections, models representing future population and land-use changes, and economic assessment tools (e.g., infrastructure costs).

Because of the limited time spent discussing these relationships, the workshops did not include a discussion of the limitations or uncertainties associated with existing H&H models and related water-quality assessment tools, although some participants highlighted this in subsequent meetings. We note that these simulation tools remain limited and imperfect using the best available current science. They can rely on simplified approaches to represent complex physical processes that may not calibrate well when compared with observed data. Similar model uncertainty could also be noted for land-use changes and climate downscaling methods, each of which adds a layer of uncertainty to this type of integrated system analysis. The uncertainty associated with these model relationships is an important current limitation but could not be addressed in the scope of this pilot effort.

Pilot Study Builds on Local Data, Tools, and Expertise

Participants in the scoping workshops identified a wide-ranging and ambitious research agenda. They collectively identified what could be considered the scope of an integrated regional watershed management planning effort, supported by uncertainty and scenario analysis, for Allegheny County watersheds. This effort encompasses the sewer overflow and water-quality planning challenge in the 83 ALCOSAN communities, but it also includes municipalities outside of the ALCOSAN service area and addresses such key goals as flood risk reduction, ecosystem management, and economic development where systematic and reliable research to support integrated planning is not currently available.

We drew from this broad scope to identify a set of discrete analysis steps that could be accomplished within the time frame and resources of this pilot study. The XLRM elements bolded in Table 2.1 represent those that we were able carry forward into the technical analysis. In developing the technical scope, we prioritized as follows:

- **Include climate change and other uncertain drivers:** Our pilot study focuses on urban response to climate change (Knopman and Lempert, 2016), and the Allegheny County region currently lacks information regarding plausible climate change effects. Therefore, including deep uncertainty of this type was a key point of focus.

- **Build on existing tools and models:** Urban watershed, stormwater, and wastewater simulation models are time- and resource-intensive to develop, calibrate, and operate. In addition, existing planning models—such as the simulation models developed for ALCOSAN's WWP—have already received careful attention and scrutiny and thus may have greater buy-in from the start. Therefore, we sought to use and adapt existing models to the greatest extent possible rather than develop new simulation modeling for the region. As described in Chapter Three and Appendix B, this entailed working with ALCOSAN's existing sewer system models.

- **Inform current deliberations:** As of this writing, negotiations regarding the draft WWP and other actions in response to the CD are still in progress. New analysis describing future vulnerabilities that could emerge or stormwater source reduction strategy performance could help inform these final decisions, and we therefore prioritized this portion of the broader stormwater scope.

Using this prioritization scheme, and informed by subsequent feedback from Study Partners and stakeholders, the RAND team translated the broader scope identified by participants into a more focused pilot effort. The resulting scope considers the performance of the sewer system across the ALCOSAN service area and focuses primarily on sewer overflows, strategy capital cost, and cost-effectiveness as key performance metrics.

Because participants identified flooding as a key concern and planning goal, we also sought to develop new estimates of climate-informed flash flood risk for the ALCOSAN municipalities. However, the RAND team was unable to develop complete or reliable modeling results for this outcome metric because of data and modeling limitations. As a result, flood risk is not described as part of the analysis results but is instead identified as an important next step. Relatedly, estimates of environmental, economic, and other co-benefits from GSI were also identified as an important planning need during the scoping workshops, but they are not yet incorporated into this pilot analysis.

Through the remainder of this study, we employ RDM with currently available simulation modeling tools. The process to translate this scope into a series of technical inputs for the analysis is briefly described, along with analysis results, in Chapters Three and Four, with more detail provided in the supporting technical appendixes (Appendixes B through E).

CHAPTER THREE

Future Sewer Overflow Vulnerability

In Chapter Two, we summarized the full regional stormwater planning scope identified in the initial workshops by study participants. These discussions were then translated into discrete technical inputs to support the remainder of the analysis, supported by a series of one-on-one conversations and follow-up with technical partners. The technical scope is summarized using the XLRM format in Table 3.1. In the next two chapters, we describe each component of this scope, in turn, describing inputs and analysis results together to the extent possible rather than detailing all inputs from the start.

In this chapter, we begin by addressing this question: How might the sewer overflow challenge grow over the next 20 to 25 years if no additional action is taken or new investments are made in the system? We begin with the sewer system as presently con-

Table 3.1
XLRM Summary of the Technical Scope

Uncertain Factors (X)	Levers and Strategies (L)

• Precipitation
  – 2003 Typical Year
  – Recent Historical (2004–2013)
  – Climate-adjusted (2038–2047)
• Temperature
  – Recent Historical (2004–2013)
  – Climate-adjusted (2038–2047)
• Impervious area (land use)
  – Current
  – Southwestern Pennsylvania Commission (SPC) growth
  – 2xPGH (high growth)
• Wastewater customers
• GSI infiltration rate
• Capital cost uncertainty
	• Policy levers and individual strategies
	  – I&I reduction
	  – GSI
	  – WWTP expansion
	  – Interceptor cleaning
	• Combined strategies

Relationships (R)	Performance Metrics (M)
• SWMM 5.1 H&H models • Downscaled climate-informed precipitation and temperature • Land-use change module (ArcGIS) • Infrastructure cost estimation tools	• CSO volume by outfall (gal.) • SSO volume by outfall (gal.) • Time in overflow by outfall (hours) • Capital cost of implementation (2016 dollars) • Cost-effectiveness ($/gal.)

structed and operated to (1) understand how the problem could worsen if policy action is delayed and (2) provide a baseline of "future without action" (FWOA) scenarios with which proposed improvements can be compared.

The chapter begins with a brief summary of the simulation models adapted for this analysis (R), followed by a description of recent historical and climate-informed precipitation and temperature inputs (X). We then provide results from the uncertainty analysis across these climate scenarios, focusing on overflow volume and time in overflow (M). Next, we describe the scenario inputs related to population growth and land use and present sewer overflow results across these scenarios in turn. All overflow scenarios are then combined to show the complete uncertainty analysis, and we highlight a handful of system outfalls that could show substantial changes in these plausible futures. The chapter concludes with a brief summary of findings and implications for future system planning and design. Chapter Four then picks up to describe the policy levers and strategies (L) formulated to test in these future scenarios. The full technical scope is represented in the analysis described in Chapter Five.

Preliminary results from the vulnerability analysis presented in this chapter were shared and discussed with Study Partners and Stakeholder Advisors in workshop settings, as discussed in Chapter Two. These discussions allowed participants to provide initial feedback on the analysis, in some cases helping to guide revisions, model reruns, or additional analysis steps, and in general helping to ensure the technical validity of the results.

Simulation Modeling to Support What-If Analysis

The RAND team adapted a series of sewer system models to support an RDM analysis for Pittsburgh and the surrounding regions. The models simulate the operations of the region's storm- and wastewater infrastructure, which enabled us to evaluate the performance of the system. Using these models and high-performance cloud computing, we simulated a ten-year sequence of overflow results in 18 different sets of assumptions about future conditions (scenarios), to (1) evaluate the potential risks and vulnerabilities to the existing storm and wastewater infrastructure resulting from long-term climate change and population growth and (2) compare the effectiveness of different strategies for reducing CSOs and SSOs across a range of climate, land-use, and population assumptions. Each unique combination of scenario and year needed to be run separately through the simulation models, yielding hundreds of separate model runs.

Given time and resource limitations, and to better compare results with recent investigations by ALCOSAN, PWSA, and other organizations, we did not independently develop the sewer system models used in this analysis. Rather, we adapted models that were developed and calibrated by ALCOSAN in the 2008–2012 time frame to develop its draft WWP. The ALCOSAN SWMM models use a full year of rainfall data to simulate the performance of the system and to estimate annual over-flow volumes and frequencies. We performed a calibration step to ensure that results from our adapted models are consistent with the original ALCOSAN results, but we did not independently alter or recalibrate the stormwater models with original flow monitoring data. As a result, *any limitations present in the original ALCOSAN models will also exist in the adapted models used in this analysis.* This is an important limitation in this context, as simulation modeling to support urban stormwater, wastewater, and water-quality planning is a challenging and evolving science. The relationships and assumptions models, such as the SWMM model, may themselves be uncertain, and simulation results may not compare well with real-world observations without recalibration and adjustment. This pilot study does not address the resulting uncertainty in the model relationships.

The process of adapting and verifying the simulation models is described briefly in the next section and in further detail in Appendix B of this report.

ALCOSAN SWMM Planning Models

ALCOSAN developed models of its member municipalities' current regional stormwater and wastewater infrastructure (as of 2012) to support development of the draft WWP using USEPA's SWMM model (ALCOSAN, 2012a, p. 1-9). USEPA developed the SWMM model to assist planners in the design and evaluation of stormwater and wastewater systems. Using a coupled H&H model, the SWMM model can simulate runoff conditions arising from single storm events or long-term precipitation scenarios. The SWMM model was originally developed in 1971, but updates have been made through many iterations to produce the current version of SWMM 5.1 (Rossman, 2015).

The ALCOSAN service area was divided into seven separate planning basins for WWP development (see Figure 2.2 in Chapter Two). ALCOSAN developed one or more SWMM models for each basin and provided RAND with a series of nine calibrated SWMM models to support this analysis. The models must be run in sequence such that the outer (upstream) basin models are run first and outputs from those models become inputs for the downstream basins. This is illustrated in Figure 3.1, where each circle represents a basin model. The final model is the Regional Balance Model (RBM), which simulates the major interceptors and trunk lines of the system.

These models evaluate the sewer system through simulations using one or more full years of rainfall data. An alternative approach that is often employed for infrastructure engineering uses discrete design storms to evaluate a system (e.g., one could design the combined sewer system to handle up to a ten-year storm event). We use continuous rainfall data rather than discrete design storms for three reasons: (1) The approach is consistent with ALCOSAN's previous analysis using the SWMM model; (2) PADEP regulations specify annual overflow reduction targets; and (3) observed rainfall data, rather than engineered design storms, better reflect the challenges facing the system.

Figure 3.1
Sequence of Basin Models Used for Automated Scripting

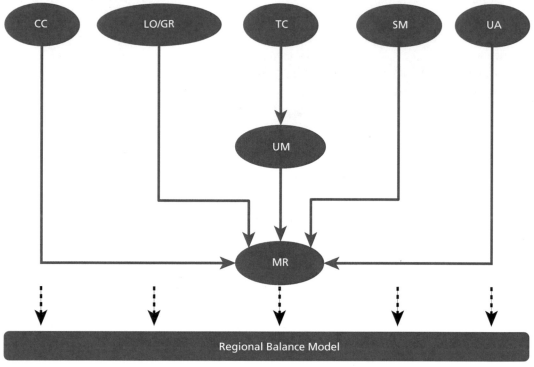

NOTE: The models must be run in sequence so that outputs from upstream basin models can be used as inputs for downstream models. LO/GR is treated as a single basin in this report but is represented in two separate SWMM models provided by ALCOSAN, yielding a total of nine simulation models.

RAND *RR1673-3.1*

Automating the Planning Models for Cloud Computing

Running all nine models in sequence to simulate a one-year period requires more than 50 central processing unit (CPU)-hours, which can be completed in roughly a 24-hour span when using a server or multicore computer that can run several basin models simultaneously. To scale the study to be able to test many variables and scenarios, the RAND team developed a computer script to automate the model sequencing using virtual servers accessed through Amazon Web Services (AWS), a cloud computing service.

The advantage of using AWS cloud computing is that the process can be scaled by running multiple simulations in parallel. Throughout the full analysis, we evaluated 585 unique scenario and year combinations in the SWMM model—including variations of rainfall and land-use scenarios and many potential strategies for reducing overflows, all discussed in greater detail later in this report. The 585 simulations required a total of approximately 30,000 CPU-hours. To put this in context, completing these simulations on a single-core desktop computer would have taken more than three years. Establishing an automated process using parallelized cloud computing allowed us to explore a more-comprehensive set of uncertainties and strategies than would have otherwise been possible.

Verifying Model Results

We went through several steps to verify that the results from our adapted models matched ALCOSAN's published results when using the same inputs and assumptions, detailed in Appendix B. For each CSO and SSO outfall, results from our adapted models were compared with the data published in ALCOSAN's draft WWP to identify any inconsistencies.

We used a newer version of the SWMM model (SWMM 5.1) in this study than the version in which the models were originally developed (SWMM 5.0) so that we could use the software to estimate GSI performance. Due to various changes between SWMM versions, rerunning the same models using the updated version resulted in slightly higher overflow volumes. A small number of outfalls accounts for much of the overflow volume discrepancy between model versions. To correct for this discrepancy, we created a bias-correction factor based on the percentage difference between our adapted SWMM 5.1 model results and ALCOSAN's published results (based on SWMM 5.0). The outputs from our adapted SWMM 5.1 models were then adjusted by this percentage difference—a bias correction—so that the overflow totals matched published results as of 2012 when using ALCOSAN's original modeling assumptions.

Table 3.2 summarizes the annual volume of sewer overflows by planning basin simulated in the RAND model after the bias correction is applied. These results closely match ALCOSAN's Existing Conditions modeled outcomes (9,509 Mgal./year total overflow).

Table 3.2
Bias-Corrected Simulated Overflows for 2003 Typical Year (Mgal.)

Basin	CSO	SSO	Total
CC	1,035	160	1,195
LO/GR	334	267	601
MR	2,836	0	2,836
SM	416	0	416
TC	149	39	188
UA	2,213	36	2,249
UM	1,956	65	2,091
Total	8,939	567	9,507

Future Overflow Uncertain Factors

The next step in our analysis was to use this modeling framework to evaluate a wider range of uncertainty than had previously been considered. Specifically, we sought to generate a wide range of possible future scenarios, which were then used to (1) evaluate vulnerabilities in the existing storm and wastewater infrastructure and (2) stress-test strategies for improving the system. In this chapter, we explore three sources of long-term uncertainty using scenario analysis:

- **Climate change:** Expanding on previous work that used rainfall from a single average or "typical" year, we created a recent historical rainfall scenario using observed data from 2004 through 2013 and then developed two climate-adjusted rainfall scenarios using projections from 2038 through 2047. The future climate projections also include air temperature increases, which can affect evapotrans-piration.
- **Land use:** We developed three land-use scenarios reflecting no population growth, moderate growth, and high growth. These scenarios changed the amount of impervious cover present in the areas of the system contributing to CSOs.
- **Wastewater customer connections:** Adapting results from ALCOSAN's WWP, we considered a scenario representing the current number of customer connections, as well as one with increased base flows and inflows resulting from an increase in the number of customer connections and an expansion of the ALCOSAN service area.

These uncertainties were identified as important and relevant for the stormwater planning challenge during the scoping workshops or had been previously identified by ALCOSAN during WWP development (customer connections). They directly affect the amount of either runoff or baseflows entering the sewer system. There is deep uncertainty in each, meaning that they are difficult or impossible to predict or reasonably weight with probability estimates over the long time horizons (25 or more years) of interest for major infrastructure projects, such as the WWP. In this and the remaining chapters of the report, each unique mapping of assumptions for these uncertain factors to consequences is referred to as a "scenario."[1] All possible combinations were considered, yielding a total of 18 scenarios simulated for the vulnerability analysis.

---

[1]   The RDM literature sometimes uses the term *future* to describe a specific realization or mapping of assumptions to consequences. This is distinguished from a "scenario" or "decision-relevant scenario," which is one outcome of an RDM analysis and can refer to a set or range of futures that share common attributes of particular relevance for a given decision or policy problem (see Lempert et al., 2013). In other words, a scenario might have an interpretable meaning, while a future is simply one combination of plausible assumptions. In this report, the study team elected not to follow this convention and uses the term *scenario* generically to describe any specific set of assumptions about uncertain factors used as inputs to the analysis.

However, note that the limited number of scenarios developed and evaluated for this vulnerability analysis do not necessarily fully capture the plausible range of uncertainty for each uncertain factor. This pilot study includes two to three different assumptions for each factor, as described next, but the number of total scenarios used to evaluate sewer overflow volumes is kept relatively low (fewer than 20) because of the substantial computing demands of the linked SWMM models. An investigation with simpler or less computationally demanding models, by contrast, might evaluate dozens or hundreds of different scenario assumptions for each factor. Currently available urban stormwater and wastewater modeling platforms do not yet support this type of large ensemble scenario analysis. As a result, the study team elected to begin with a relatively small scenario set for this pilot investigation.

In the following section, we present results with climate uncertainty first, then consider customer and land-use changes, and finally incorporate all scenarios together through this chapter.

Rainfall and Climate Change Uncertainty

Recent studies have begun to investigate the effect that climate change may have on combined sewer systems in cities around the world in terms of drainage system performance (Kleidorfer et al., 2009; Semadeni-Davies et al., 2008), flood risks (Nilsen et al., 2011), and sewer overflow volumes (USEPA, 2008). Using the "Model for Urban Sewers" hydraulic modeling software, for example, two separate studies (Nilsen et al., 2011; Semadeni-Davies et al., 2008) found that predicted precipitation patterns estimated from climate models could greatly exacerbate the issues faced in the Nordic cities of Helsingborg, Sweden, and Oslo, Norway, both of which experience CSOs.

However, methods to incorporate climate change into sewer system planning or design are relatively new and have not yet been widely adopted. As a result, many urban areas—including Pittsburgh and ALCOSAN municipalities—have not yet been able to consider long-term climate change effects on their systems.

Recent Local Investigations Rely on a "Typical Year" Approach

Allegheny County's weather is influenced from systems moving from the Gulf of Mexico, Canada, and the U.S. Central Plains. Mean yearly precipitation in the county ranges upward of 35 to 40 inches per year, with about 60 percent occurring as rainfall during the spring and summer seasons. The many hills and valleys of the region can lead to wide spatial variations in stormwater runoff and rainfall infiltration (Michael Baker Jr., Inc., 2014, pp. 12–13).

ALCOSAN developed a "typical rainfall year" in drafting the WWP, an approach described in USEPA guidance and commonly applied to support LTCP development (USEPA, Office of Wastewater Management, 1995, 1999). After examining the 60-year historical rain gauge data (1948 through 2008) from the Pittsburgh International Airport, the WWP team identified a long-term average of 36.7 inches per year and found that precipitation from the 2003 rainfall year best matched the historical average at the time of analysis. The rainfall inputs into the SWMM models are based on radar-adjusted rainfall observations at a 15-minute time step and a 1-km^2 spatial resolution. The high-resolution rainfall data were, in some cases, adjusted so that event statistics better matched the historical airport data (ALCOSAN, 2012c, p. 4-3). We refer to this adjusted 2003 rainfall data as the *2003 Typical Year* throughout this report. Since draft WWP development, local planners have used the 2003 Typical Year to represent or characterize regional hydrology (e.g., PWSA, 2016).

New Scenarios Reflect Recent or Plausible Future Changes

As noted in Chapter One, the most recent summaries of projected climate effects suggest that the northeastern United States is expected to see increased precipitation in a warming climate. This could include more precipitation overall (a higher annual average), a higher number of severe rainfall events (greater intensity of storms), or both (Melillo, Richmond, and Yohe, 2014; Shortle et al., 2015). Building on the 2003 Typical Year, we identified or developed three additional hydrology scenarios to test, representing either the recent hydrology since 2003 or two projections of plausible future climate effects. These scenarios are summarized below; a full description of the climate downscaling methods applied is provided in Appendix C of this report. As noted previously, these scenarios are plausible, but they do not necessarily fully capture the range of uncertainty associated with future rainfall patterns.

Recent Historical hydrology

The first hydrology scenario (Recent Historical) builds on ten years of high-resolution radar-adjusted rainfall observations throughout Allegheny County recorded and maintained by 3 Rivers Wet Weather (undated-a). In this case, the 1 km^2 of observed precipitation data was reformatted and used directly as inputs for the SWMM models for the years 2004 to 2013.

Climate Downscaled Hydrology

Additional climate scenarios were developed using a process called "downscaling." In this approach, we combined the high spatial and temporal resolution from the recent observed data with the longer-term, low-resolution precipitation trends projected from global and regional climate models. There are many possible downscaling methods that have been applied in different regions and contexts. Here, we applied a method called the "non-parametric delta-change method," in which we calculated a change factor, or delta, from the difference between future and historical projections, and then applied the change to observed data (Arnbjerg-Nielsen et al., 2013; Boe et al., 2007; Wilks and Wilby, 1999; Wood et al., 2004). The advantage of this approach is its relative simplicity compared with other approaches and that it does not rely on underlying assumptions about the probability distribution of the data (Gudmundsson et al., 2012). An important limitation, however, is that it assumes that the frequency and duration of storms remain the same. With this approach, only the magnitude (intensity) of rainfall is changed.

Future climate projections were obtained from Regional Climate Model (RCM) outputs from the North American Regional Climate Change Assessment Program (NARCCAP) (Mearns et al., 2009). NARCCAP provides a compilation of RCMs that have been forced by General Circulation Models (GCMs), or RCM-GCM combinations (Mearns, 2014). RCMs estimate climate outputs at a higher spatial and temporal resolution than global models and can support more-detailed assessments of climate effects, such as urban stormwater analysis. The projections utilize the Special Report on Emissions Scenarios (SRES) A2 greenhouse gas emission scenario for the 21st century, one of the highest-emission scenarios during previous iterations of global climate analysis from 2000 to 2007 (Nakicenovic et al., 2000).

Out of the 11 different RCM-GCM combinations available from NARCCAP, we then identified two specific model combinations to represent the range of plausible outcomes emerging from the simulations. The RCMs were selected based on their performance in a "hindcast" simulation for Allegheny County—in other words, running the model over a historical period and then comparing the results with actual observed statistics. The Higher Intensity Rainfall scenario uses the HRM3-GFDL[2] model combination to represent a future with the highest-intensity daily storms, yet a marginal increase in total annual rainfall. In this scenario, storms will get much more severe, but dry periods between them will get longer. The Higher Total Rainfall scenario, alternatively, uses the MM5I-HadCM3[3] simulation. This scenario projects the largest increase in total annual rainfall and slightly less-intense daily storms than the Higher Intensity Rainfall scenario. It represents a future with a similar frequency of rain events as the past, but more rain during these events.

Using the methods described in Appendix C, we downscaled precipitation results from these two models to represent the ten-year period from 2038 to 2047, approximately 25 to 30 years into the future. Because changes to air temperature can also influence the SWMM hydrology simulation, we also updated the temperature inputs for the SWMM model when running these scenarios using values from the corresponding RCM-GCM combination for this future period at a one-day time step and the 50-km^2 grid. Temperature outputs for these models were identified for the grid cell closest to the centroid of Allegheny County, and no additional spatial or temporal downscaling was conducted for air temperature.

---

[2] The Hadley Regional Model 3 (HRM3) forced by the Geophysical Fluid Dynamics Laboratory General Circulation Model (GFDL).

[3] Penn State University/National Center for Atmospheric Research (MM5I-PSU/NCAR) Mesoscale Regional Climate Model (MM5I) forced by the Hadley Center Coupled Model, version 3 (HadCM3).

Although the selected climate model combinations performed well in hindcast simulations for the region, they are but two of many projections of future precipitation developed in recent years. They are based on a single assumption about the pathway of future global carbon emissions (or emission scenario; see Appendix C) and represent a small subset of the many global climate change simulations evaluated over the past decade. Furthermore, not all climate models agree on the direction or trend of future precipitation for different regions of the United States, and various simulations may show positive (wetter), negative (drier), or no trend for the Northeast (Melillo, Richmond, and Yohe, 2014; van Oldenborgh et al., 2013). The climate scenarios used here are generally consistent with the ensemble average across climate models for the northeastern United States, as noted in Chapter One, but the uncertainty between models remains high. This uncertainty is exacerbated when applying downscaling methods, which add assumptions and simplifications.

As a result, these scenarios should be considered an initial set of plausible stressing futures for Allegheny County and the ALCOSAN sewer system, but they do not represent the full range of current climate model projections. The scenario development described here is a novel and significant first step for the region, but additional scenarios might yield different patterns in terms of average rainfall or frequency of intense storms, and it is too soon to tell which of these futures might be more or less likely to occur.

Final Climate Scenarios

The resulting outputs from the downscaling analysis are two climate-adjusted precipitation scenarios, at a 15-minute time step and 1-km^2 spatial resolution, from 2038 to 2047, along with corresponding temperature inputs. Table 3.3 shows an example of the summary statistics for the resulting precipitation at a point in the middle of the system (MR basin). Included in the left pane are average annual rainfall totals for each of the three scenarios described earlier, including the mean and the range across the ten-year sequences. On the right pane, we look at the frequency of more-intense rainfall and count the number of days with more than trace rainfall or higher than selected rainfall thresholds.

Overflow Results from Recent Historical (2004–2013) Simulations

Table 3.3 shows several interesting patterns. First, the Recent Historical rainfall for MR is about an inch higher than the 2003 Typical Year, on average, and ranges from 30 to 50 inches across the ten-year sequence. Over the past decade, the basin has had about 80 days per year with more than trace precipitation. About six days per year had more than 1 inch of rainfall, and, about one day per year, local rainfall exceeded 2 inches. In the two downscaled climate scenarios, however, both the average and the extremes increase. In the Higher Intensity Rainfall scenario, for instance, the annual average is

Table 3.3
Summary Statistics for Three Precipitation Scenarios (Main Rivers Planning Basin)

Climate Scenario	Years	Annual Rainfall (inches)			Average Days Per Year with Rainfall Above Threshold			
		Min	Mean	Max	0.1 in.	1 in.	1.5 in.	2 in.
Recent Historical	2004–2013	29.8	38.9	50.0	80.5	5.9	2.2	0.9
Higher Intensity Rainfall	2038–2047	34.3	40.9	52.0	81.2	6.9	2.6	1.4
Higher Total Rainfall	2038–2047	36.5	43.1	53.9	84.7	8	3.0	1.7

NOTE: The 2003 Typical Year annual rainfall totaled approximately 37.8 inches for the MR basin, according to our statistical summary. This is about an inch higher than the average identified by ALCOSAN at the airport gauge.

2 inches higher than in Recent Historical, and it rains 1 inch or more another day per year on average. Higher Total Rainfall shows further extremes, with an average annual rainfall increase of more than 4 inches and greater frequency of rainfall extremes.

These statistics suggest an increase in annual rainfall volumes in Allegheny County over the past decade when compared with the previous 60 years. Another look at the long-term data collected at Pittsburgh International Airport is consistent with this increase and shows some evidence that an upward trend in average rainfall is already occurring. Specifically, Figure 3.2 shows the annual rainfall totals at the airport gauge from 1953 to 2015, including a simple linear trend fitted to the data. The linear fit is influenced by very high rainfall years in 1990 and 2004 but nevertheless appears to show a gradual upward trend in average rainfall over this period.

The annual rainfall average appears to increase from 35 inches at midcentury to closer to 38 to 40 inches at present. The 2003 Typical Year was developed based on the 1948–2008 average and without the benefit of the most recent available rainfall years (2009 to 2015) and thus may not fully capture the recent change.

We next considered whether these increases might also be affecting sewer overflows. Using the adapted modeling system described earlier, we evaluated annual overflows from the 2003 Typical Year and the subsequent decade (2004 to 2013) (Recent Historical), holding wastewater customer connections and land use at their current (2012) levels. Figure 3.3 provides a scatterplot summary of the resulting overflow volumes. CSOs are shown on the y-axis and SSOs on the x-axis, with each point representing one year of simulation. Points are sized according to total overflow volume, and the guidelines highlight the 2003 Typical Year results for comparison.

Figure 3.3 shows that most years in the Recent Historical scenario yield higher CSO, SSO, and total overflow volumes than in the 2003 Typical Year simulation. Only 2009 yields lower CSO and SSO volume, while 2005 and 2008 show higher

Figure 3.2
Annual Rainfall at Pittsburgh International Airport, 1953–2015 (Inches)

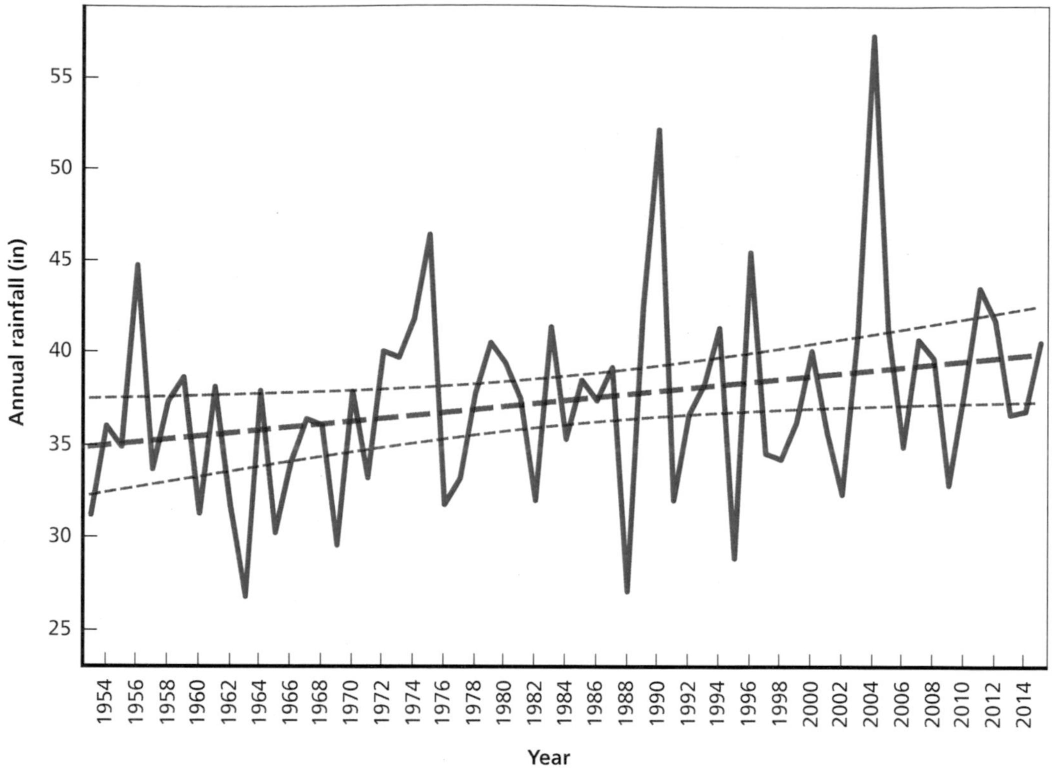

SOURCE: National Centers for Environmental Information, 2016.
NOTE: The trend line shows a simple linear fit to annual average rainfall, while the bounding lines show the 95-percent confidence interval around this estimate. The fitted annual trend is 0.08 inches/year (0.01, 0.15) and is statistically significant at the 95-percent confidence level.
RAND RR1673-3.2

SSOs but not CSOs. For all other years (upper right quadrant), the CSO and SSO volumes are higher, and total overflow exceed 13 Bgal./year in 2004 and 2011.

Total overflow volumes in recent years are correlated with rainfall, as expected, but the relationship differs between hydrology assumptions. Figure 3.4, for example, shows total monthly overflows compared with a simple average of systemwide rainfall volume by month. Each point represents one month in either the 2003 Typical Year (dark gray) or Recent Historical (light gray) climate scenario, and a separate trend line is included for each.

The correlation is evident in Figure 3.4 (R^2 of 0.70–0.78), with a steeper slope for 2004 through 2013 than for the 2003 Typical Year. This suggests that, even in months with similar rainfall volumes, the overflow volumes can be somewhat higher when considering the full range of years from 2004 through 2013. In turn, this provides evidence that the pattern of storms—that is, the location, timing, and intensity

Figure 3.3
Simulated Annual CSO and SSO Volumes, 2003 Typical Year and Recent Historical

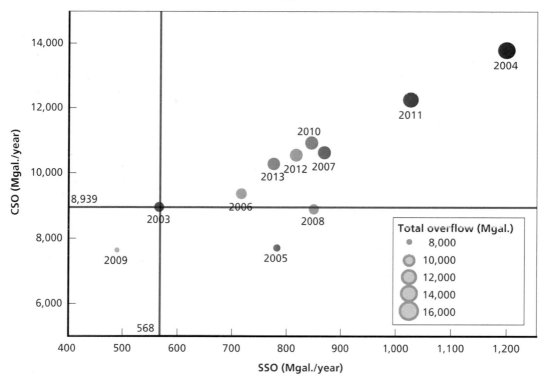

NOTE: The vertical and horizontal gray lines show the SSO and CSO volumes for 2003 Typical Year rainfall, respectively.

RAND *RR1673-3.3*

of rainfall across the basin—in the Recent Historical set may differ from the adjusted 2003 values, and this, in turn, may yield higher overflows than anticipated.

Figure 3.5 illustrates this further by looking at the seasonal pattern. Here, we summarize total overflow by month using just the 2003 Typical Year (dark gray line) or an average across the ten-year Recent Historical scenario (light gray line). The 2003 Typical Year simulation shows lower overflows in the spring months and a higher summer peak, while the more recent ten-year average shows higher springtime totals and a more-consistent pattern of flows throughout the year.

Table 3.4 summarizes these results by planning basin, comparing the 2003 Typical Year simulation with a ten-year annual average from Recent Historical. All basins show higher overflows in the ten-year simulation when compared with the typical year results, ranging from a 7- to 38-percent increase. Average annual overflows in Recent Historical are simulated at 11 Bgal./year, an increase of approximately 16 percent compared with previous estimates.

Figure 3.4
Rainfall and Overflows by Month, 2003 Typical Year and Recent Historical

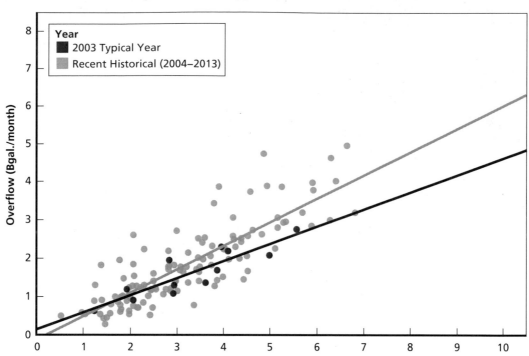

NOTE: Monthly rainfall volumes represent the unweighted average across all basins. 2003 Typical Year values are highlighted. A single outlier with high rainfall and overflows (September 2004, which included Hurricanes Ivan and Frances) is omitted. The correlation coefficient (R^2) equals 0.78 for the 2003 Typical Year and 0.70 for the 2004–2013 period.

RAND RR1673-3.4

Figure 3.6 provides a mapped summary of Recent Historical overflows by outfall. Each point is sized according to its average annual overflow, with colors showing the time in overflow per year (in hours). The map confirms that most of the outfalls contributing high overflow volumes and overflowing frequently throughout the year are located immediately along the Allegheny and Monongahela rivers and close to the metropolitan center, with particular challenges along the Allegheny. The CC basin also shows several outfalls with substantial time in overflow.

Overflow Results in Future Climate Simulations

We next added simulations of future climate uncertainty to these results, again using a ten-year average (2038 through 2047) to better look across year-to-year variability. As a starting point, these results hold constant customer connections and land use at cur-

Figure 3.5
Comparison of Average Monthly Overflow, 2003 Typical Year and Recent Historical

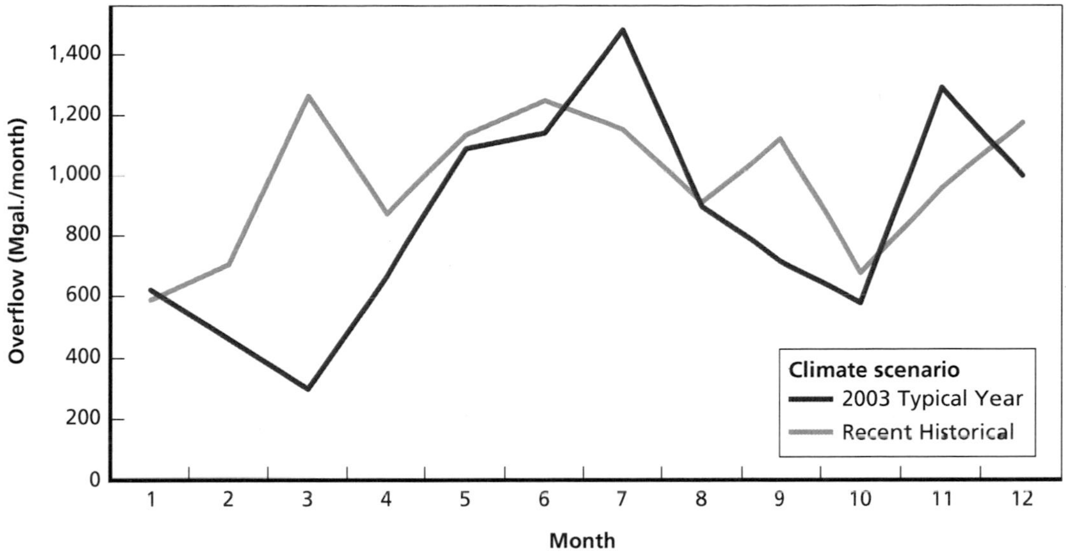

RAND RR1673-3.5

Table 3.4
Recent Historical Simulated Overflows Compared with 2003 Typical Year

Basin	2003 Typical Year	Recent Historical Average	Percentage Change
CC	1,195	1,414	18
LO/GR	601	797	33
MR	2,836	3,032	7
SM	416	574	38
TC	188	222	18
UA	2,249	2,567	14
UM	2,022	2,433	20
Total	9,507	11,039	16

rent values to focus just on climate effects. Overflow volume results are summarized in Figure 3.7 for each climate scenario, compared with the 2003 Typical Year. Shades of gray in the stacked bar plot show the separate contributions of SSOs and CSOs, respectively, to total overflow volume.

Both projected climate scenarios, as expected, yield higher SSO and CSO volumes than were simulated in the 2003 Typical Year or Recent Historical hydrology,

Figure 3.6
Map of Recent Historical Overflows by Outfall

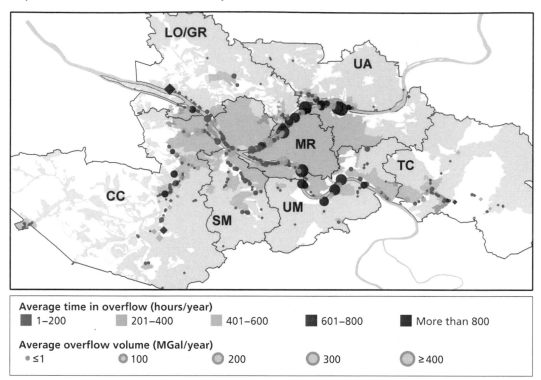

NOTE: Circles represent CSO outfalls; diamonds show SSO outfalls. Selected outfalls, including interceptor relief overflow (IRO) outfalls identified by ALCOSAN, are omitted for clarity.
RAND RR1673-3.6

with Higher Total Rainfall producing the highest total overflow volume (12.1 Bgal./year). Most of this change occurs in CSO rather than SSO volumes. But the magnitude of change is relatively modest when compared with Recent Historical, with increases of 600 Mgal./year (5 percent) or 1.1 Bgal./year (10 percent) for the Higher Intensity Rainfall and Higher Total Rainfall scenarios, respectively. The 1.5-Bgal./year difference between Recent Historical and the 2003 Typical Year is greater than either of these.

One possible explanation is that the difference in storm patterns between the 2003 Typical Year and the ten-year Recent Historical sequence has a more substantial effect on overflow volumes than the increase in rainfall intensity reflected in the downscaled climate projections. In other words, the system may be more sensitive to changes in storm frequency or distribution across the region; because the relatively simple downscaling approach keeps the Recent Historical pattern of storms fixed, the estimated climate effect appears relatively modest. This suggests that future climate downscaling efforts for the region may need to consider alternative storm patterns to better estimate plausible future vulnerability.

Figure 3.7
Overflow Volumes with Climate Uncertainty

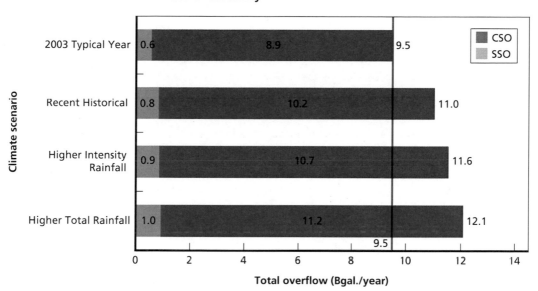

NOTE: The vertical gray line shows the estimated total overflow using Current Connections, Current Land Use, and the 2003 Typical Year rainfall for visual comparison purposes.

RAND RR1673-3.7

Future Population and Land-Use Uncertain Factors

In considering future vulnerability for the regional sewer system, the RAND team also developed scenarios representing plausible future population and land-use uncertainty. The general goal was to represent a range of plausible futures for the region. Nevertheless, a limited set of scenarios was developed and simulated in the SWMM model as a starting point given high computation costs. Similar to the climate scenarios, these population and land-use scenarios therefore may not describe the full range of uncertainty, and they represent a first step toward incorporating deep uncertainty into regional sewer system planning.

Next, we briefly describe how each scenario dimension was developed. A full description of the methods and resulting scenarios can be found in Appendix C.

New Wastewater Customers

Our modeling results are based on ALCOSAN's Existing Conditions SWMM model, which represents the system as it existed in 2012. The Existing Conditions model assumes an estimated 836,600 customers residing within the 214 square miles of service area. ALCOSAN developed a separate Future Baseline model to represent expected growth by 2046. The Future Baseline model assumes a modest increase in the number of serviced customers, up to 969,000, and an expansion of the service area in the CC and TC basins resulting in a total footprint of 233 square miles.

To account for a potential increase in future customer connections, we created a new scenario dimension for wastewater customer connections using a simple post-processing step to adjust results from the Existing Conditions model to those that would be expected under ALCOSAN's Future Baseline model conditions. This was done as an alternative to running the Future Baseline models directly, which was prohibitive when combined with other scenario dimensions because of computing costs.

As described in Appendix C, we developed a fixed additive term at the outfall level based on a comparison of ALCOSAN modeled results and then applied the same additive term to every outfall when estimating overflow volumes with future customer connections. This makes the conservative assumption that increases in wastewater customer flow will have an approximately linear effect on overflows even when other changes occur in other rainfall and land-use scenarios. Current Connections represents the current number of customers; Future Connections yields results with this added flow included. In our simulations, this change results in approximately 500 Mgal./year of added overflows, about 75 percent of which (364 Mgal./year) are increased CSO volumes (see Table C.2 in Appendix C).

Changing Patterns of Land Use

Another key driver for stormwater runoff into urban combined sewer systems is the extent to which land area is covered by impervious surfaces such as buildings, roads, and parking lots. This driver could interact with changing rainfall patterns to drive future vulnerability—Kleidorfer et al. (2009), for example, suggests that impervious area reduction may be a key requirement to offsetting the increase in runoff expected from increased rainfall intensities.

Recognizing that (1) the extent of impervious cover in the ALCOSAN combined service area is likely to change in the future as the region's population and economic development patterns evolve and (2) both population and development patterns are deeply uncertain when looking out several decades or more, we developed an approach to create plausible land-use scenarios to support an investigation of future change and vulnerability to additional sewer overflows. For this analysis, we developed land-use scenarios to represent three plausible future conditions: Current Land Use, which assumes no change in population or impervious cover, a moderate-growth scenario (SPC Growth), and a high-growth scenario (2xPGH).

Land-use scenarios were developed through a two-step process. First, we identified possible future populations roughly 30 years from the present day, spanning a wide range. Second, we used a method derived from the peer-reviewed literature (Exum et al., 2005; see Appendix C for additional explanation) to estimate the change in impervious cover resulting from changes in population density for the combined sewer area (i.e., the area of the ALCOSAN system that contributes stormwater into to the combined sewer system; see Figure 2.2 in Chapter Two).

In this analysis, future land use is treated as an uncertainty and not included as a modeled policy lever in later chapters. However, the RAND team notes that enforcement of existing or new stormwater ordinances or other land-use regulations could help to limit or avoid increases in impervious cover in Allegheny County municipalities, essentially reducing the likelihood of such scenarios as SPC Growth or 2xPGH.

Population Growth Projections

The first step in creating the land-use scenarios was to identify plausible projections of future population in the ALCOSAN combined sewer area. For the moderate-growth scenario, we used a recent population projection developed by the SPC, which projects approximately 15-percent growth by 2046 (see Appendix C). This is the same starting point used by ALCOSAN and its basin planners to estimate future customer connections. We refer to this as the SPC Growth scenario. To test a fairly extreme upper limit, by contrast, we also generated a separate high-growth scenario (2xPGH), in which the population in Pittsburgh nearly doubles by 2046, equivalent to a 2-percent annual increase. This also roughly corresponds to the city's historical population peak. Given Pittsburgh's recent history of population decline or flat growth, this scenario may, at present, seem unrealistic. However, over the span of the next three decades, national and regional population trends are sufficiently uncertain that a reversal of the historical population trend or sudden spike in growth remains plausible, and this scenario is therefore a useful bounding case.

Table 3.5 summarizes the resulting population change by 2046; note that these are combined sewer area populations only (see Figure 2.2).

Impervious Cover and Final Land-Use Scenarios

Next, we estimated the change in impervious cover for the combined sewer area—represented in the SWMM models directly connected impervious area (DCIA)—corresponding to each population projection using the methods described in Appendix C. From a baseline of 7,711 acres DCIA in current conditions (Current Land Use), this approach yielded DCIA increases of 8 and 24 percent, respectively, for the SPC Growth and 2xPGH scenarios. Table 3.6 summarizes the final projected DCIA by planning basin for each of the three land-use scenarios, and Figure 3.8 shows an example of DCIA change by subcatchment for the SPC Growth scenario.

Overflow Results with Population and Land-Use Uncertainty

We next estimated overflows in these customer connection and land-use scenarios, holding hydrology constant using the Recent Historical scenario. A summary bar plot is shown in Figure 3.9, in the same format as Figure 3.7.

The first row shows the same results as before, with Current Connections and Current Land Use yielding 11 Bgal./year of overflows. As noted previously, moving

Table 3.5
Population Projections for Each Land-Use Scenario, by Basin, Combined Sewer Area

Planning Basin	ALCOSAN Service Population (2010)	SPC Growth Projected (2046)		2xPGH Projected (2046)	
		Population	Percentage Difference	Population	Percentage Difference
CC	29,833	31,071	4	40,330	36
LO/GR	13,583	15,054	11	17,941	32
MR	142,663	169,262	19	281,731	97
SM	29,492	32,797	11	55,606	89
TC	9,488	11,318	19	11,318	19
UA	42,663	47,188	11	75,116	76
UM	25,276	30,662	21	41,169	63
Total	292,998[a]	337,352	15	523,211	79

[a] Totals are slightly different from ALCOSAN WWP estimates because of methodology differences and changes in the geospatial file.

Table 3.6
DCIA for Land-Use Scenarios by Planning Basin

Planning Basin	Current Land Use Acres	SPC Growth (2046)		2xPGH (2046)	
		Acres	Percentage Difference	Acres	Percentage Difference
CC	657	712	8	723	10
LO/GR	306	321	5	335	9
MR	3,900	4,251	9	5,300	36
SM	429	466	9	535	25
TC	119	139	17	144	21
UA	1,432	1,507	5	1541	8
UM	866	944	9	948	9
Total	7,709[a]	8,340	8	9,526	24

[a] Totals are slightly different from ALCOSAN WWP because of methodology differences and changes in the geospatial file.

Figure 3.8
Change in DCIA by 2046 (SPC Growth Scenario)

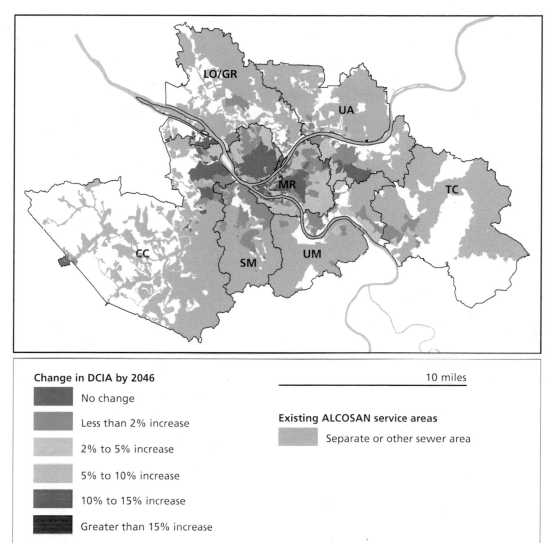

from Current Connections to Future Connections adds 500 Mgal./year without accounting for land use. When we consider impervious cover increases in the combined sewer area, however, we see that a DCIA increase of 8 percent yields another approximately 500 Mgal./year of overflows; adding 24 percent, alternatively, adds 1.2 Bgal./year compared with Current Land Use. As expected, increased DCIA influences CSO volumes mostly through the additional stormwater runoff from paved surfaces in the combined sewer area.

Figure 3.9
Overflow Results with Population and Land-Use Uncertainty (Recent Historical Rainfall)

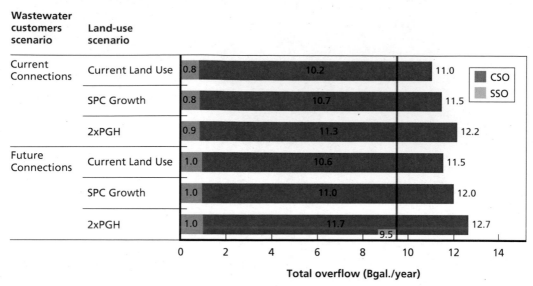

NOTE: The vertical gray line shows the estimated total overflow using Current Connections, Current Land Use, and the 2003 Typical Year rainfall for visual comparison purposes.

RAND RR1673-3.9

The plausible range of increases is notable but somewhat surprising. The 2xPGH scenario represents a relatively extreme worst-case scenario, one in which the urban population doubles and impervious surface in the urban core increases by nearly 25 percent. But even in this relatively extreme scenario, total overflow increase by only 10 percent compared with Current Land Use. By contrast, a DCIA increase of 8 percent (SPC Growth) yields 4.5 percent more overflows, with a larger ratio of DCIA to flow increase. This suggests that the system is less vulnerable to future population growth in the combined area. In other words, although the system faces substantial challenges under present conditions, the combined area appears to already be substantially "built out" and is not as exposed to future growth. As a consequence, land-use planners in the urban core may choose to focus more on addressing existing land uses and pavement than on preventing future impervious cover growth.

Overflow Results from All Simulated Scenarios

The final step in this analysis of vulnerability was to simulate all combinations of the scenarios described earlier. Average annual overflow volumes are shown in Figure 3.10, using the same format as prior results.

When combining customer connections, land-use, and climate scenarios together in a future with no additional action, the range of average annual overflows extends

Figure 3.10
Overflow Results from All Scenarios Considered

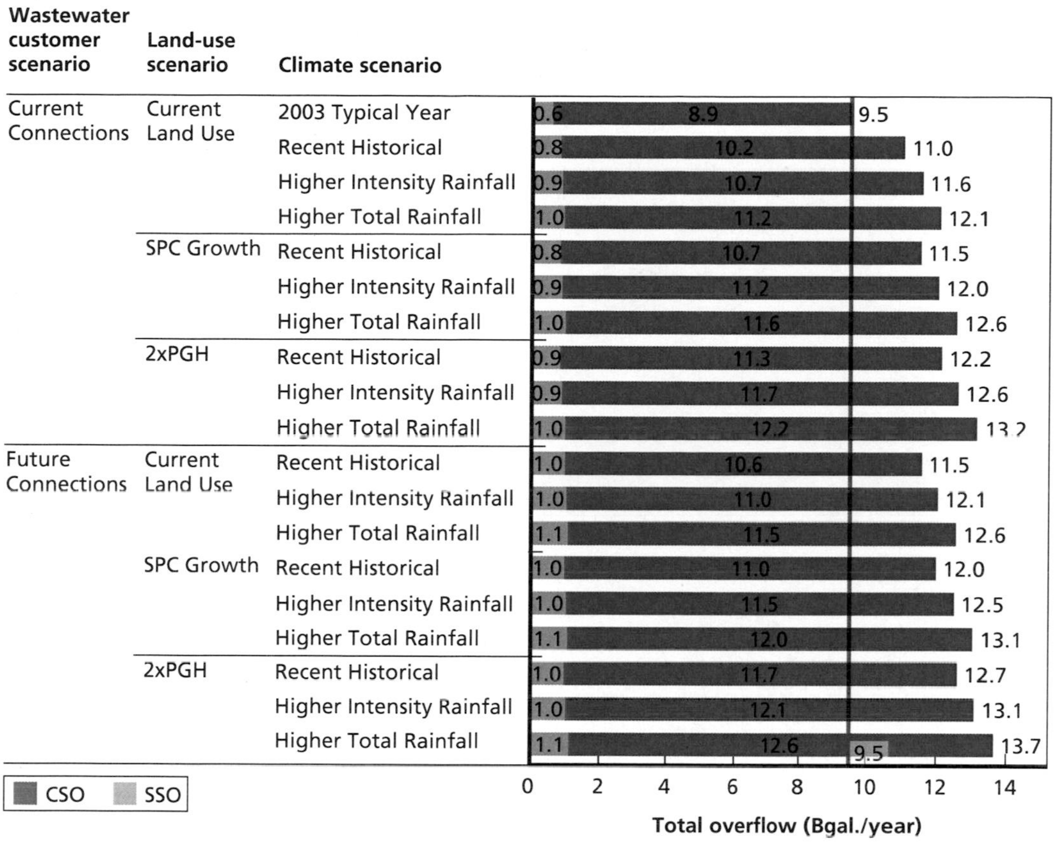

Wastewater customer scenario	Land-use scenario	Climate scenario	Total overflow (Bgal./year)
Current Connections	Current Land Use	2003 Typical Year	0.6 / 8.9 / 9.5
		Recent Historical	0.8 / 10.2 / 11.0
		Higher Intensity Rainfall	0.9 / 10.7 / 11.6
		Higher Total Rainfall	1.0 / 11.2 / 12.1
	SPC Growth	Recent Historical	0.8 / 10.7 / 11.5
		Higher Intensity Rainfall	0.9 / 11.2 / 12.0
		Higher Total Rainfall	1.0 / 11.6 / 12.6
	2xPGH	Recent Historical	0.9 / 11.3 / 12.2
		Higher Intensity Rainfall	0.9 / 11.7 / 12.6
		Higher Total Rainfall	1.0 / 12.2 / 13.2
Future Connections	Current Land Use	Recent Historical	1.0 / 10.6 / 11.5
		Higher Intensity Rainfall	1.0 / 11.0 / 12.1
		Higher Total Rainfall	1.1 / 11.5 / 12.6
	SPC Growth	Recent Historical	1.0 / 11.0 / 12.0
		Higher Intensity Rainfall	1.0 / 11.5 / 12.5
		Higher Total Rainfall	1.1 / 12.0 / 13.1
	2xPGH	Recent Historical	1.0 / 11.7 / 12.7
		Higher Intensity Rainfall	1.0 / 12.1 / 13.1
		Higher Total Rainfall	1.1 / 12.6 / 13.7

CSO ■   SSO ▨

NOTE: The vertical gray line shows the estimated total overflow using Current Connections, Current Land Use, and the 2003 Typical Year rainfall for visual comparison purposes.

RAND RR1673-3.10

from 11 to 13.7 Bgal./year (16 to 44 percent greater than the current system with 2003 Typical Year hydrology, top row). SSO volumes stay relatively constant—ranging from 0.8 to 1.1 Bgal./year—while CSO volumes can increase by 2 Bgal./year or more, depending on the scenario. Looking at a middle-range scenario—Future Connections/SPC Growth/Higher Intensity Rainfall, for instance—overflows increase to 12.5 Bgal./year, or 32 percent more than previously estimated. These results undergird that, across a range of plausible scenario assumptions, the overflow challenge is changing and is likely to grow absent additional investments in the system.

Another consideration is the reliability of reducing or eliminating overflows. The previous results have shown a ten-year average, summarizing across a range of years in terms of storm patterns and intensity. But if ALCOSAN and municipal planners seek to reliably eliminate overflows, this would entail having sufficient capacity to address high rainfall and high flow years, not simply the average.

Figure 3.11 provides an illustration of this challenge, showing results across all land-use and climate scenarios with future customer connections with box plot summaries across all ten years in the sequence.

When considering all of the simulated years, it is evident that, even if improvements to the sewer system were to reliably eliminate average flows, high-rainfall years would likely produce overflows. This is significant from both a reliability and regulatory perspective, because eliminating SSOs in an average or typical year only may not constitute full compliance. Looking at Higher Intensity Rainfall and SPC Growth (middle pane, middle box), for instance, we see a wider range than suggested by the 12.5 Bgal./year average. The 75th percentile is 13 Bgal./year, and the extreme range extends up to above 18 Bgal./year. Similar ranges are observed in all other scenarios considered. If ALCOSAN and the municipalities seek to eliminate overflows in 75 percent of all years, this would yield a higher target for stormwater and wastewater strat-

Figure 3.11
Total Overflow Box Plot Across Ten-Year Simulation Period

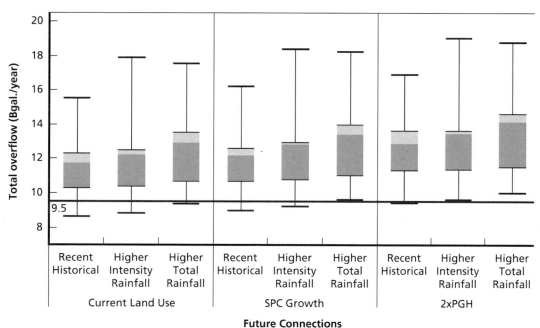

NOTE: Future Connections scenario shown. The box plots show the 50th percentile (median; horizontal line where the two gray shaded areas meet), interquartile range (IQR) (edges of each box), and maximum extent across each ten-year sequence (whiskers). The 2003 Typical Year results are again shown with a reference line for comparison.

RAND RR1673-3.11

egy design than the average; a 90-percent or 95-percent threshold would be yet higher, targeting overflows in the 14- to 16-Bgal. range.

Scenario Results by Outfall

The scenario results described earlier show plausible overflow changes likely to affect the entire sewer system. But greater change will occur in some areas of the system than others, with some outfalls exhibiting much higher increases in overflow volume or time spent in overflow than others. Figure 3.12 illustrates these differences by showing the change occurring by outfall in a single scenario combination—Future Connections/ SPC Growth/Higher Intensity Rainfall. For convenience, we will use this scenario for selected results through the remainder of this chapter.

Figure 3.12 shows the change occurring at each outfall between the baseline (Current Connections/Current Land Use/Recent Historical) and this selected scenario.

The mapped results show that many outfalls with small overflows—for example, outfalls in the TC basin in the southeast portion of the system—may show relatively high proportional increases in plausible future scenarios. That said, most outfalls contributing substantial overflow volumes also show change, the magnitude of which likely contributes much or most of the topline changes observed.

Table 3.7 shows a wider range of results for the 12 outfalls highlighted in Figure 3.12, including overflow volumes, time in overflow (hours per year), and the proportional increases in both outcomes for the selected scenario. Compared with the baseline, overflow volumes at these outfalls generally increase 13 to 34 percent. The future scenario shown yields increases in overflow, in some cases exceeding 100 Mgal./year at a single outfall (e.g., M-29-OF). Time in overflow changes range from relatively constant to increases of up to 23 percent. One CSO outfall, municipal outfall 2000-774 in the CC basin, shows much higher proportional increases, but this seems to be an outlier among the highest-volume outfalls.

Finally, to further illustrate the plausible range of changes in detail, we show results across all scenarios for one outfall and associated sewershed area that has a major point of focus for stormwater planning for Pittsburgh and PWSA: A-22. The A-22-OF outfall is connected to a sewershed that includes the Pittsburgh neighborhoods of Shadyside, Oakland, Bloomfield, Friendship, Garfield, and Polish Hill. It appears at the top of the Table 3.7 list in terms of overflow volumes, and changes in this area would be of interest for local GSI planning efforts occurring concurrently in this area (e.g., PWSA, 2016).

Results for A-22 range fairly widely across the scenarios considered. A-22 outfall overflow volumes are similar between the 2003 Typical Year and Recent Historical climate scenarios (Table 3.8, top two rows; 3-percent difference), although the time in overflow appears to take a step upward when considering 2004 through 2013 instead

Figure 3.12
Map of Overflow Change for One Future Scenario

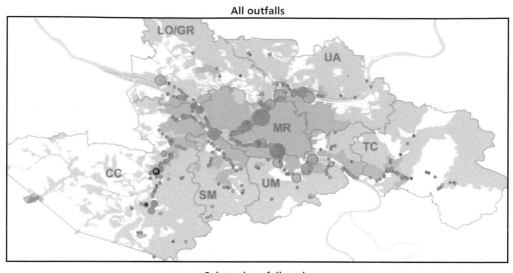

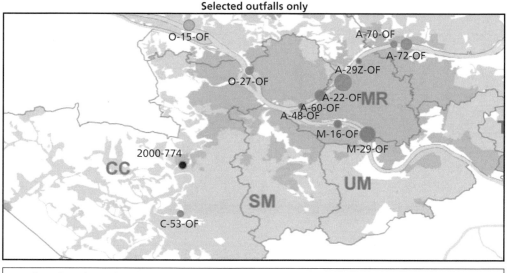

NOTE: Shows scenario with Future Connections, SPC Growth land use, and Higher Intensity Rainfall. Change is calculated relative to Current Connections, Current Land Use, and Recent Historical Rainfall. Some outfalls are omitted for clarity, including IRO outfalls. Point sizes show the average annual overflow volume, while colors illustrate the percentage change in overflow volume relative to the baseline. Larger points with darker red shades indicate outfalls likely to disproportionately contribute to future overflows. The top pane shows all outfalls, while the bottom pane provides a zoom and highlights 12 outfalls contributing a substantial fraction of total overflows.

Table 3.7
Results from Two Scenarios for Selected Outfalls

| Outfall | Basin | Type | Current Connections, Current Land Use, and Recent Historical | | Future Connections, SPC Growth, and Higher Intensity Rainfall | | | |
			Overflow (Mgal./ year)	Overflow Time (hours/ year)	Overflow (Mgal./ year)	Percentage Difference	Overflow Time (hours/ year)	Percentage Difference
A-22-OF	MR	CSO	610	644	686	12	657	2
M-29-OF	MR	CSO	472	843	579	23	880	4
O-15-OF	LO/GR	SSO	245	943	254	4	975	3
A-72-OF	UA	CSO	238	399	268	13	401	1
A-60-OF	MR	CSO	221	501	270	22	471	−6
O-27-OF	MR	CSO	113	234	131	16	258	10
M-16-OF	MR	CSO	101	475	128	27	583	23
A-70-OF	UA	CSO	83	2,279	96	16	2,530	11
C-53-OF	CC	SSO	81	664	96	19	760	14
2000-774	CC	CSO	68	868	152	124	1,677	93
A-48-OF	MR	CSO	56	90	75	34	101	12
A-29Z-OF	MR	CSO	55	338	73	33	366	8

of 2003 alone. Otherwise, overflow volumes increase modestly with additional rainfall intensity (8 to 9 percent) across the climate scenarios, with little further change noted for overflow hours.

Future customer connections do not meaningfully change results in this sewershed, largely because ALCOSAN does not project notable new customer growth in this already highly developed area. In terms of land use, the SPC Growth scenario yields a proportionally small change when holding other factors constant (Table 3.8, fourth row; 4-percent increase with Current Connections and Recent Historical climate). Greater effects are noted only with the more extreme 2xPGH scenario, which would yield about a 20-percent increase in overflow with no further climate effect (Table 3.8, eighth row). In general, the range of plausible overflow volume increases in A-22 across scenarios is 0 to 192 Mgal./year (0 to 32 percent) compared with Recent Historical, with 0 to 66 additional hours in overflow per year (0 to 10 percent). A range of different patterns emerges, however, when looking at scenario results for outfalls not described in detail here.

Table 3.8
Detailed Results from All Scenarios for Outfall A-22-OF

Wastewater Customers Scenario	Land-Use Scenario	Climate Scenario	Overflow (Mgal./ year)	Percentage Difference	Overflow Time (hours/ year)	Percentage Difference
Current Connections	Current Land Use	2003 Typical Year	593	−3	538	−16
		Recent Historical	610	0	644	0
		Higher Intensity Rainfall	656	8	648	1
		Higher Total Rainfall	665	9	669	4
	SPC Growth	Recent Historical	635	4	653	1
		Higher Intensity Rainfall	682	12	657	2
		Higher Total Rainfall	699	15	683	6
	2xPGH	Recent Historical	728	19	683	6
		Higher Intensity Rainfall	778	28	684	6
		Higher Total Rainfall	799	31	710	10
Future Connections	Current Land Use	Recent Historical	613	1	644	0
		Higher Intensity Rainfall	659	8	648	1
		Higher Total Rainfall	668	10	669	4
	SPC Growth	Recent Historical	639	5	653	1
		Higher Intensity Rainfall	686	13	657	2
		Higher Total Rainfall	703	15	683	6
	2xPGH	Recent Historical	731	20	683	6
		Higher Intensity Rainfall	782	28	684	6
		Higher Total Rainfall	802	32	710	10

NOTE: Percentage change is calculated relative to Current Connections, Current Land Use, and Recent Historical (2004–2013) climate.

Conclusion

In this chapter, we introduced the framework and methods used to conduct a pilot study of future sewer overflow vulnerability for municipalities in the ALCOSAN service area. Using a set of existing simulation models developed by ALCOSAN to support WWP development, we conducted a series of quantitative experiments using high-performance computing to explore how the system as currently constructed and operated might respond to plausible future scenarios reflecting climate, wastewater customer connections, and land-use changes.

The results of this exercise suggest that the overflow challenge may have already increased given rainfall patterns over the past decade and could grow further with plausible future change. The extent of increased vulnerability depends on the assumptions made but ranges from 1.5 to 4.2 Bgal./year in additional overflow volume when compared with a 2003 Typical Year simulation. We caution, however, that these numbers are based on a limited number of uncertain factors and scenarios, constrained by available computing resources. Including additional uncertain factors or additional plausible futures within the factors could yield a wider range of outcomes.

Working versions of this analysis were presented to and discussed with Study Partners and Stakeholder Advisors during workshops in winter and spring 2016. Participants provided constructive feedback on the initial scenarios developed, modeling assumptions, and results, all of which helped strengthen the final analysis. In general terms, participants accepted the validity of the simulation modeling approach and subsequent vulnerability analysis results. Detailed partner and stakeholder feedback from these workshops was gathered through a formal and separate evaluation process, and results from this evaluation will be published in a forthcoming report.

All of these results are based on an FWOA in which the current system is maintained as it is, in large part to provide a baseline against which future investments can be compared. In Chapter Four, we identify and test a range of proposed planning-level strategies intended to improve system function or reduce the flow of stormwater into the system during rainfall events. In Chapter Five, we then evaluate a promising subset of these strategies across the scenarios introduced here, also reflecting uncertainty in terms of strategy cost and GSI performance.

Comparing Source Reduction and Wastewater Policy Levers

In the previous chapter, we explored long-term vulnerabilities for Allegheny County's storm and wastewater infrastructure, assuming no improvements or capital investments into the existing system. In this chapter, we develop and test 30 strategies aimed at reducing overflows, building off of a core set of policy levers identified as relevant for the region. Key policy levers include variations of broadly applied GSI, reductions in I&I through pipe and manhole repairs, upgrading the capacity of the treatment plant, and removing debris from the main interceptors to increase conveyance capacity.

In this chapter, we briefly describe each of the policy levers, along with the technical inputs and modeling assumptions. We present results showing the reduction in overflows resulting from individual levers, as well as "combination" strategies, which assess the value of pursuing multiple approaches simultaneously. Finally, this chapter introduces first-order cost estimates for each strategy.

We refer to this chapter as our *screening strategy* analysis; the goal was to provide a preliminary assessment to help select a final set of strategies for the RDM analysis, which is described in Chapter Five. The results presented here are based on a one-year simulation using rainfall from the 2003 Typical Year, which we use to provide more-consistent comparisons with previous analyses by ALCOSAN and others. Cost values presented in this chapter include only single point estimates. While these are useful for this preliminary screening analysis, they should not be used to identify the best-performing or most cost-effective strategy. As emphasized throughout this report, uncertainty cannot be ignored. The point estimates presented here are intended to provide preliminary insights and a starting point for analyses in Chapter Five, in which we explore the effects of various strategies under a wide range of possible future conditions reflecting uncertainties in both performance and cost.

Policy Levers Considered in This Pilot Study

Next, we briefly introduce each policy lever evaluated in this pilot analysis. These levers include key source reduction options proposed for the region, as well as several proposed near-term improvements to the existing sewer system. The levers initially evaluated were identified through a combination of feedback from Study Partners and Stakeholder Advisors during participatory workshops and sensitivity testing conducted by the modeling team.

However, there were various policy levers not evaluated in this pilot study because of time and resource limitations. These included most of the sewer system infrastructure improvements proposed by ALCOSAN in its draft WWP, such as new conveyance pipes or deep interceptor tunnels (ALCOSAN, 2012g); strategies targeting stream daylighting or groundwater inflows (discussed further in Appendix D); or sewer separation in the combined areas, which is likely to be cost-prohibitive because of the challenge of installing a new stormwater pipe network alongside the sewer system in already developed and densely populated areas. Further information regarding model approach, assumptions, and data sources can be found in Appendix D.

Green Stormwater Infrastructure

GSI encompasses a wide range of technologies and approaches for managing stormwater runoff, including rain barrels, rain gardens, bioretention, infiltration trenches, and green roofs (3 Rivers Wet Weather, undated-c). There are two mechanisms by which GSI can reduce overflows. First, GSI can remove runoff—either through infiltration into groundwater or evapotranspiration—that would otherwise enter a combined sewer system. Second, GSI can act as distributed storage that can hold runoff during a rainfall event and slowly release it back into the combined sewer system. By delaying the release, GSI may increase the chance that the system will have adequate capacity to handle the stored runoff without contributing to overflows (3 Rivers Wet Weather, undated-f).

In this analysis, we consider a set of high-level GSI strategies, similar to those evaluated in ALCOSAN's recent source control study (ALCOSAN, 2015b, Chapter Three). These strategies are applied using simplified criteria to the entire ALCOSAN combined sewer area, represent first-order approximations, and do not take into account site-level characteristics, constraints, or other key barriers to implementation (see Appendix D for more information).

We developed five strategies assuming that GSI will be installed in each subcatchment of the combined sewer area. All GSI was modeled as bioretention in the SWMM model. Table 4.1 shows the key assumptions for each strategy, which include the following:

- **DCIA controlled:** A measure of the scale of a GSI strategy or installation. The four GSI strategies are sized to capture runoff from 1 inch or 1.5 inches of rainfall over 10, 20, or 40 percent of total DCIA in the combined sewer area.
- **Loading ratio:** The ratio of the tributary area of DCIA to the GSI footprint. Results presented in this chapter assume a loading ratio of 10:1, where each acre of GSI captures runoff from 10 acres of DCIA, and GSI is sized for 10 inch-acres of runoff volume (or 15 inch-acres if the strategy is sized for 1.5 inches of rainfall). Results from a sensitivity analysis (Appendix D) show a significant perfor-

Table 4.1
GSI Strategy Assumptions

Strategy Name	GSI Type	GSI Sizing	Infiltration Rate (Inches per Hour)	Loading Ratio
GSI-10	Bioretention	10% of DCIA with 1″ rain	0.1	10:1 and 25:1
GSI-20	Bioretention	20% of DCIA with 1″ rain	0.1	10:1 and 25:1
GSI-40	Bioretention	40% of DCIA with 1″ rain	0.1	10:1 and 25:1
GSI-40-HI (High Infiltration)	Bioretention	40% of DCIA with 1″ rain	0.2	10:1 and 25:1
GSI-40-AOP (Art of the Possible)	Bioretention	40% of DCIA with 1.5″ rain	0.2	10:1 and 25:1

NOTE: Strategy names refer to the percentage of DCIA used to determine the size of the GSI. For example, GSI-20 is sized according to the volume of runoff from 1 inch of rain over 20 percent of DCIA in the combined sewer area. All results presented in this chapter assume a 10:1 loading ratio; results presented in Chapter Five include loading ratios of both 10:1 and 25:1.

mance improvement with "high-flow" sites (i.e., high loading ratio). Because of the importance of this assumption, in the RDM analysis presented in Chapter Five, we include loading ratios of both 10:1 and 25:1, corresponding to "Low" or "High" GSI performance uncertainty. In other words, 1 acre of GSI will capture runoff from either 10 or 25 acres of DCIA. Note that we keep the size of the GSI fixed; in High GSI cases, we are simply adjusting the acreage of tributary area routed to the GSI installation.

- **Infiltration rate:** The rate at which GSI infiltrates stormwater into the ground, thereby removing it from the system. Three of the five GSI strategies assume an infiltration rate of 0.1 inch per hour, and the remaining two strategies assume 0.2 inch per hour.

A sensitivity analysis of GSI performance parameters is included in Appendix D. Table 4.2 shows the total storage volume for each GSI strategy, the GSI footprint, and the tributary runoff area assuming either a 10:1 or 25:1 loading ratio.

Note that there are two different types of infiltration discussed in this chapter. Infiltration from GSI into groundwater can eliminate runoff that would otherwise enter the combined sewer system, potentially reducing overflows. In the context of the I&I reduction strategies discussed in the following section, by contrast, infiltration occurs when groundwater or rainfall seeps into cracked or broken sewer pipes, potentially increasing overflows. In short, infiltration from GSI can be beneficial, while pipe infiltration has a negative effect on the sewer system.

Also note that the most aggressive GSI targets, which are sized to control runoff from 1 inch or 1.5 inches of rainfall for 40 percent of DCIA across the entire com-

Table 4.2
Total GSI Installed, by Strategy

Strategy Name	Total GSI Storage Volume (Mgal.)	Total GSI Footprint (Acres)	Tributary DCIA Runoff Area; 10:1 Loading Ratio (Acres)	Tributary DCIA Runoff Area; 25:1 Loading Ratio (Acres)
GSI-10	26	97	971	2,427
GSI-20	53	194	1,943	4,857
GSI-40	106	389	3,886	9,715
GSI-40-HI	106	389	3,886	9,715
GSI-40-AOP	158	389	3,886	9,715

bined sewer area, may or may not be realistic. Such an aggressive strategy would likely need to deploy a variety of technologies beyond bioretention alone, and all installations would be subject to slope, soil, and other engineering conditions. These strategies would require up to 389 acres dedicated to new GSI, much of which would need to be in or near the most densely populated areas. To put this into context, 389 acres of GSI are equivalent to roughly 300 football fields (including end zones). As another point of local reference, Pittsburgh's Schenley Park is 456 acres and Frick Park is 644 acres.

For further context, we reviewed existing green infrastructure plans in other cities, finding a wide range of targets, a mix of implementation success, and some optimism about future technological and process improvements.[1] At one extreme, the Milwaukee Metropolitan Sewerage District announced in 2013 that it had a target of up to 42,000 acres of GSI by 2035, accounting for more than 70 percent of its current impervious area—although it has yet to report on its initial results (Milwaukee Metropolitan Sewerage District, 2013). By contrast, New York City set a goal in 2010 of managing 10 percent of its approximately 14,000 impervious acres with green infrastructure by 2030. The city recently fell short of its five-year milestone of 1.5 percent, with completed projects controlling only 0.6 percent of impervious area (New York City Department of Environmental Protection, undated).

Philadelphia also identified an ambitious target and has seen early success with implementation. The city proposed creating approximately 9,600 "greened acres" (impervious area controlled) across 34 percent of the city over the next 25 years. In the first five years of its Green City, Clean Waters program, Philadelphia exceeded its goal of 744 acres of GSI by more than 25 percent, reducing overflows by 1.5 Bgal. with 441 stormwater infrastructure projects (Philadelphia Water Department, 2013). As a

---

[1]   Cities use varying terminology that limits apples-to-apples comparisons. For example, Milwaukee's ambitious target includes nonengineered solutions on private property (e.g., an individual replacing a home parking pad with lawn), whereas another city might only include designed "infrastructure," such as bioretention swales on public property.

final example, Washington, D.C., recently included green infrastructure in its updated wastewater LTCP. In two basins that were originally targeted for gray infrastructure, Washington, D.C., plans to use GSI to control nearly 500 impervious acres to manage stormwater by 2030 (District of Columbia Water and Sewer Authority, 2015). In the coming years, cities with substantial investments in GSI will likely report on their progress, and their findings could inform future planning in Allegheny County.

The more ambitious GSI targets included in this analysis were developed to explore the upper bound of GSI's potential contribution to reducing overflows, recognizing that experience and technology improvements could make previously infeasible approaches more realistic in future decades. If the region chooses to pursue a large-scale GSI strategy, however, further engineering and design work, as well as site-based implementation criteria informed by ownership and maintenance, would be needed to identify a feasible and realistic target.

Pipe Repair to Reduce Inflows and Infiltration

I&I from rainfall and groundwater inflows (GWI) result in a high volume of water entering the regional sewer system—approximately 47 Bgal./year, or more than half of the volume treated at the Woods Run plant (ALCOSAN, 2015b). By taking up valuable capacity in the pipe network, I&I can contribute to overflows. As previously mentioned, *infiltration* in this context should not be confused with GSI infiltration; the former increases the volume of water in the sewer system and contributes to overflows, while the latter helps to reduce overflows by eliminating runoff that would otherwise enter the system.

I&I is the result of an aging sewer system with leaks in manholes, customer laterals, municipal sewers, and major trunk lines and interceptors, as well as some cases in which buried streams flow directly into the system. The process for reducing I&I typically involves a flow isolation study to identify target areas, repairing or relining pipes, and sealing manholes with watertight frames and covers. There is significant uncertainty about the level of I&I reduction that would result from a regional pipe or manhole repair effort. The strategies developed and tested here were largely informed by two recent studies from the Pittsburgh region, which are discussed in Appendix D.

We developed six possible I&I strategies based on two different target areas and three levels of reduction that may be achieved. A selection criteria was used to identify target areas with high rainfall-derived I&I (RDII), the primary I&I contributor to overflows. The criterion is based on the average R-value for each subcatchment. An R-value is a measure of the rainfall that enters the sewer system through I&I, where higher values indicate areas with higher infiltration from rainfall into the pipe network. Figure 4.1 shows the two resulting target areas; note that they are in only the separate sewer areas (i.e., areas of the system with separate stormwater and wastewater pipes). The levels of assumed I&I reductions are shown in Table 4.3, where R1, R2, and R3 are the short-, medium- and longer-term RDII responses (see Appendix D).

Figure 4.1
Target Areas Selected for I&I Reduction Based on RDII Criteria

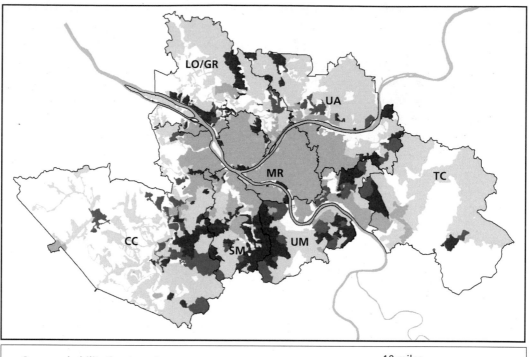

Sewer rehabilitation target areas

R-value: 6 to 8%

R-value: 8% or greater

10 miles

Existing ALCOSAN service areas

Combined sewer area

Separate sewer area

Non-contributing runoff to combined

NOTE: An R-value is a measure of the rainfall that enters the sewer system through I&I; higher values indicate areas with higher infiltration into pipes from rainfall.
RAND RR1673-4.1

Table 4.3
Three Levels of I&I Reduction Resulting from Pipe and Manhole Repairs in Target Areas

Assumed Reduction	RDII Reduction	GWI Reduction (%)
Low	10% reduction in R1 20% reduction in R2 and R3	10
Mid	20% reduction in R1 40% reduction in R2 and R3	20
High	40% reduction in R1 60% reduction in R2 and R3	30

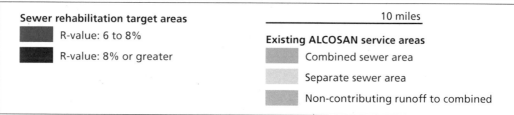

Treatment Plant Expansion

As noted in Chapter Two, USEPA has requested an expansion of ALCOSAN's Woods Run WWTP as an early step toward WWP implementation, and this project has been initiated. In this policy lever, we simulate an expansion of the treatment plant from its current capacity of 250 MGD to an expanded capacity of 480 MGD. See Appendix D for additional information.

Deep-Tunnel Interceptor Cleaning

This lever simulates the effects of cleaning the debris from the existing main interceptor tunnels to increase their conveyance capacity. The main interceptor tunnels run along the Allegheny, Monongahela, and Ohio rivers, carrying stormwater and wastewater from the ALCOSAN service area to the main treatment plant. These tunnels make up approximately 30 of the 92 miles of pipes that ALCOSAN maintains throughout the service area. Currently, gravel and other sediment accumulates as wet weather events wash debris into the sewer system. The accumulation of material creates choke points along the main interceptors, limiting their capacity and potentially contributing to overflows. See Appendix D for additional information.

Preliminary Strategy Cost Assumptions

Drawing on previous work, we estimate the expected range of capital costs for each strategy. For the screening analysis in this chapter, we present initial point estimates for total capital costs; in the next chapter, we incorporate a cost range, which reflects the significant cost uncertainty associated with each of these levers. Next, we briefly summarize the cost assumptions used in this analysis, with further discussion in Appendix D.

This is a preliminary, first-order cost analysis. Values used throughout this report include only the estimated capital cost for the strategies; we do not consider the cost of ongoing operations and maintenance (O&M) or the expected service life of a project, both of which are important for accurately assessing the full "life-cycle" cost of a strategy. O&M costs were excluded because of the added complexity, additional assumptions needed, and lack of supportive data needed to incorporate them into this pilot investigation. Excluding these factors (as we do here) may bias these results to favor strategies with low capital cost, even if they have higher lifetime costs from high O&M or short life spans. In addition, we do not consider financing costs, which could be significant for major infrastructure investments. The full cost burden of any strategy may be significantly higher than the capital costs presented here.

Finally, the affordability and feasibility of a strategy will depend on who pays and how. For example, GSI could be implemented by individual homeowners—encouraged through subsidies, incentives, or requirements—or it could be implemented through large-scale projects constructed by municipalities, ALCOSAN, or a future stormwater authority. The research presented here is a first step at comparing costs and benefits across fundamentally different approaches for addressing sewer overflows. Further work is needed to estimate life-cycle costs in more detail and to develop a financing and implementation strategy.

GSI Capital Costs

Table 4.4 shows the assumed GSI capital costs used in this analysis. The GSI costs are based on values from the Alternatives Costing Tool (ACT), which was used by Philadelphia Water and adapted by ALCOSAN for cost estimates in each of their stormwater management plans. We use direct construction costs assuming improved development practices and economies of scale directly from the ACT (Philadelphia Water Department, 2009, Table 2.3.1-6). Construction costs are assumed to be "retrofit" rather than "redevelopment."[2] Direct construction costs are adjusted for location and inflation, and we add adjustments to account for non-construction costs (see Appendix D). Note that the total capital cost also includes engineering and design, materials and installation, contractor's profit and overhead, and a construction contingency; the cost values do not include the value of land.

GSI was represented in the SWMM models as bioretention. However, for costing purposes we assume that GSI could be a mix of the different control technologies and that the performance (in terms of reducing overflows) would be comparable to bioretention, as modeled in the SWMM model. For the initial screening, we assume a nominal mix of GSI in which 95 percent would be the lower cost alternatives (bioretention, porous pavement, or subsurface infiltration) and 5 percent would be higher-cost green roofs.

I&I Reduction Capital Costs

Table 4.4 also shows the I&I cost assumptions used in this analysis. The pipe repair costs are based on cost data from actual repair efforts in the region (ALCOSAN, 2015a), and manhole repair costs were adopted directly from the Philadelphia ACT (Philadelphia Water Department, 2009). We nominally assume that 40 percent of the pipes and manholes in a target area will need repairs to achieve the target I&I reduction. This assumption is an important cost driver, however, which we vary and explore in greater detail in Chapter Five.

---

[2]   According to the ACT, retrofits are roughly 30 percent more costly because redevelopment includes only the marginal cost of construction when redevelopment work is already taking place (Philadelphia Water Department, 2009).

Table 4.4
Capital Cost Assumptions for Screening Analysis (2016 Dollars)

Cost Area	Units	Nominal Amount
GSI		
Bioretention, porous pavement, subsurface infiltration	$/impervious area	$285,000
Green roof	$/impervious area	$672,000
I&I		
Pipe repair cost	$/linear foot	$144
Manhole repair cost	$/manhole	$2,500
Treatment plant expansion and interceptor cleaning		
Plant expansion	Millions of $	$335
Interceptor cleaning	Millions of $	$200

NOTE: GSI values are based on the cost per impervious acre controlled, not the cost of an acre of GSI installed.

Treatment Plant Expansion Capital Costs

We assume a nominal cost of $335 million (2016 dollars) to expand the treatment plant to 480 MGD from its current capacity of 250 MGD. This value is adopted directly from the ALCOSAN WWP (ALCOSAN, 2012e). Along with the increased main pump capacity, this accounts for expanding systems for secondary treatment, as well as on-site conveyance and disinfection.

Interceptor Cleaning Capital Costs

While there is significant uncertainty about the feasibility and cost of cleaning the deep-tunnel interceptor, for this analysis, we adopt a preliminary cost estimate provided by PWSA of approximately $200 million (PWSA, 2016). This estimate is intended to include both the cost of new drop shafts that are needed to get machinery into the tunnels and the cost of the cleaning itself.

Results from Screening Comparisons

Overflow Reduction

The screening strategies provide a preliminary evaluation of the performance of various stormwater reduction strategies and near-term infrastructure upgrades aimed at reducing CSOs and SSOs. Screening strategies are evaluated using a single scenario and year assumption: Current Wastewater Customer Connections, Current Land Use, and 2003 Typical Year rainfall. The naming convention for the strategies is as follows:

- **FWOA:** Future Without Action is the baseline for the strategy screening analysis, which is based on the Existing Conditions model and the average-year precipitation (modified in 2003). All screening strategies are evaluated relative to the FWOA model results.
- **GSI:** There are five variations of GSI; the assumptions and parameters of those strategies are listed in Table 4.1. The strategy names refer to the percentage of DCIA used to determine the size of the GSI installations. For example, the GSI-20 strategy is sized to capture runoff from 1 inch of rainfall over 20 percent of DCIA in the combined sewer area. Note that, for the screening strategies, we assume a loading ratio of 10:1 (e.g., each acre of GSI collects runoff from 10 acres of DCIA). For the final RDM analysis presented in Chapter Five, we test loading ratios of 10:1 and 25:1 to represent GSI performance uncertainty.
- **I&I:** The *I&I* label refers to pipe repair strategies aimed at reducing RDII and GWI, as discussed in Appendix D. The strategy is defined by the expected level of I&I reduction (low, mid, or high) and the target area for pipe repair (areas with R-values greater than 6 percent or 8 percent). For example, I&I High (6 percent) is the most aggressive I&I strategy, which targets areas with R-values greater than 6 percent and for which the repairs yield a high level of I&I reduction (40-percent to 60-percent reduction in RDII and 30-percent reduction in GWI, as discussed in Appendix D).
- **Treatment plant expansion:** The label *480 MGD* refers to a treatment plant expansion to a daily capacity of 480 MGD, up from the current capacity of 250 MGD.
- **Interceptor cleaning:** The label *Clean* refers to cleaning the main interceptors to increase conveyance capacity.

In total, we evaluated 30 strategies in this phase. Figure 4.2 presents results from the screening strategy simulations, where the x-axis shows the total system overflows for the 2003 Typical Year. Strategies are clustered and color-coded based on the lever type; the number to the right of each bar gives the annual change in overflows, in Bgal. per year, relative to a FWOA (9.5 Bgal./year). Similarly, Figure 4.3 shows the average time in overflow, including both CSO and SSO outfalls, for each screening strategy. This metric provides a high-level measure of the frequency and duration of overflows across the system. For each strategy, the average time in overflow was calculated as the hours per year that an overflow was occurring, averaged across all outfalls in the system. We compared strategy results with the FWOA, with an average time in overflow of 188 hours. Results show the following:

- GSI strategies (Strategies 1 through 5), which are sized to capture runoff from 1 inch to 1.5 inches of rainfall over 10 to 40 percent of DCIA in the combined

Figure 4.2
Simulated Overflow Reduction with Screening Strategy (2003 Typical Year)

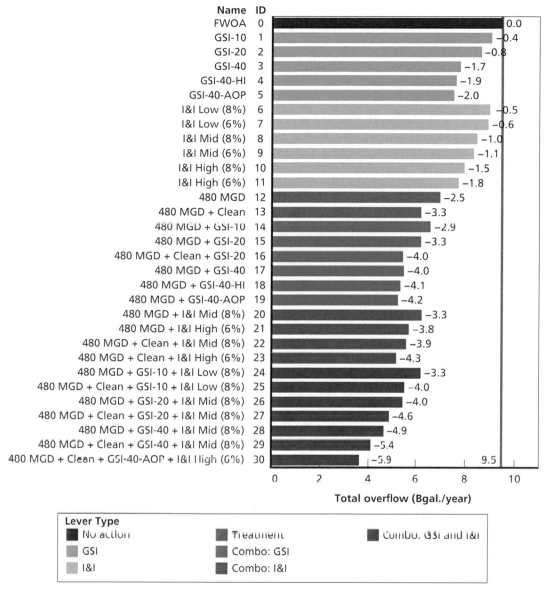

NOTE: Gray vertical line = overflow, in Bgal./year, in FWOA. The number to the right of each bar gives the annual change in overflow, in Bgal./year, relative to an FWOA (9.5 Bgal./year).

RAND RR1673-4.2

Figure 4.3
Average Time in Overflow per Outfall with Screening Strategy (2003 Typical Year)

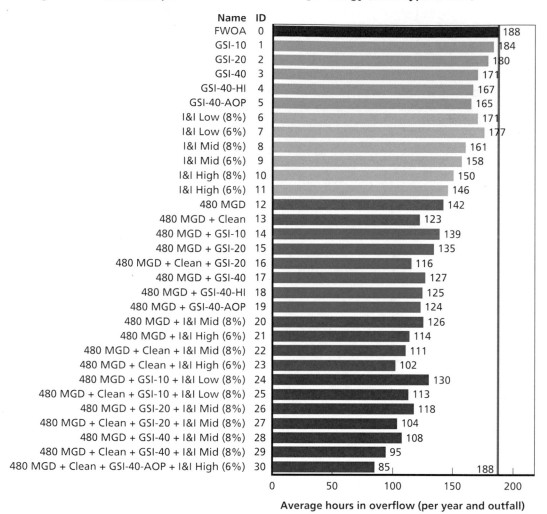

NOTE: Gray vertical line = average hours in overflow in an FWOA.

RAND RR1673-4.3

sewer area, reduce overflows 440 Mgal. to 2 Bgal./year (reductions of 5 to 21 percent from FWOA) and reduce the average time in overflow 2 to 12 percent.

- I&I strategies in isolation (Strategies 6 through 11) reduce overflows 500 Mgal. to 1.8 Bgal./year (reductions of 5 to 19 percent) and reduce the average time in overflow 9 to 22 percent.

- Expanding the treatment plant is expected to reduce overflows by 2.5 Bgal./year (26-percent reduction), and expanding the treatment plant and removing debris from the main interceptors reduce overflows by 3.3 Bgal./year (a reduction of 34 percent). These strategies reduce the average time in overflow by 24 and 35 percent, respectively. Note that Strategies 14 through 30 all include an expansion in the treatment plant, reflecting the fact that this project is already moving forward with implementation. Strategies 14 through 19 show results from strategies that combine the treatment plant expansion with GSI, yielding overflow reductions of 2.9 to 4.2 Bgal./year (31- to 44-percent reduction) and reductions in the average time in overflow of 26 to 34 percent.

- The performance of Strategy 16 (480 MGD + Clean + GSI-20) is roughly equal to Strategy 17 (480 MGD + GSI-40), with overflow reductions of 4 Bgal./year (42-percent reduction). This suggests two different pathways for achieving similar outcomes: a very aggressive GSI strategy (GSI-40) or a less aggressive GSI strategy (GSI-20) paired with cleaning the main interceptors.

- Strategies 20 and 21 show results for strategies that combine treatment plant expansion with I&I, resulting in 3.3 to 3.8 Bgal./year in overflow reductions (35 to 40 percent) and a 33- to 39-percent reduction in the average time in overflow; adding the interceptor cleaning (Strategies 22 and 23) yields overflow reductions of 3.9 to 4.3 Bgal./year (41- to 45-percent reduction) and a 41- to 46-percent reduction in the average time in overflow. Strategies 24 through 30 test combinations across all lever types with increasingly aggressive variations of GSI and I&I. Results show overflow reductions ranging from 3.3 to 5.9 Bgal./year (35- to 62-percent reduction) and a reduction in the average time in overflow of 31 to 55 percent.

Note that results are not necessarily additive when implementing multiple lever types together. For example, in isolation, the treatment plant expansion and GSI-40-HI result in 2.5 and 1.9 Bgal./year in reductions, respectively. However, combining the two levers into a single strategy (Strategy 18) results in 4.1 Bgal./year in reductions—roughly 7 percent less than the total of the two strategies in isolation. This is a case of diminishing returns: After a first strategy is implemented and overflow reductions are achieved, further strategies may be less effective because the "low-hanging fruit" has already been addressed and further reductions are harder to achieve. This effect is more pronounced with more aggressive strategies. For example, Strategy 30 is a combination of three strategies: I&I High (6 percent), GSI-40-AOP, and the treatment plant expansion, along with cleaning the main interceptors. The combined strategy reduces 5.9 Bgal./year in overflows—17 percent less than the total of the three strategies in isolation.

Figure 4.4 shows results for the same set of 30 screening strategies, where the x-axis and y-axis are the total system CSO and SSO overflows, respectively (note that the axes are on different scales). Again, strategies can be compared with an FWOA, which has roughly 9 Bgal./year in CSOs and 570 Mgal./year in SSOs. GSI strategies (green; Strategies 1 through 5), which were implemented only in the combined sewer area, primarily reduce CSOs. I&I strategies (blue; Strategies 6 through 11), by contrast, target the separated sewer system and, as a result, have a much greater impact on SSOs.

Figure 4.4
CSO and SSO Results with Screening Strategy (2003 Typical Year)

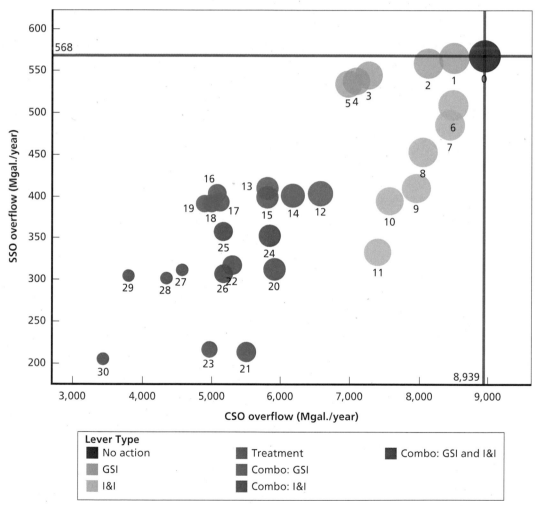

NOTE: Points are sized according to the remaining total overflow volume, and the guidelines highlight the FWOA results for comparison.

RAND RR1673-4.4

The treatment plant (Strategy 12) offers a more balanced effect on CSOs and SSOs, with a 26-percent reduction in the former and a 29-percent reduction in the latter. Note that the most aggressive strategy (Strategy 30) eliminates roughly 62 percent of CSOs and 64 percent of SSOs.

None of the strategies explored here completely eliminates SSOs, a requirement of the PADEP CD. There are two important factors that limit strategy performance for SSO reductions. First, GSI is not an appropriate solution in separated areas; we assume GSI would be installed only in the combined sewer areas, and results show a limited effect on SSOs. Second, the treatment plant expansion and cleaning the main interceptors significantly reduced SSOs in the LO basin but had a negligible effect on SSOs in the SM, CC, TC, UM, and GR basins. This suggests that the outlying sewer systems lack the necessary conveyance capacity to benefit from an expanded treatment plant.

Capital Costs and Cost-Effectiveness Results

Figure 4.5 shows the annual reduction in overflows (y-axis) and the total strategy capital cost (x-axis). This comparison was used to identify a promising subset of strategies for the full uncertainty analysis, and a star indicates those strategies carried forward for further analysis in Chapter Five.

Across the 30 strategies, nominal costs range from roughly $300 million to $2.7 billion, with annual overflow reductions ranging from 440 Mgal. to 5.8 Bgal. The figure provides preliminary insights into the cost-effectiveness of the various strategies, with points closest to the lower left corner achieving the greatest overflow reduction at the lowest cost. Results show the following:

- **GSI (Strategies 1 through 5):** Assuming $304,000 per impervious acre controlled,[3] the total cost of GSI strategies that control 10 to 40 percent of DCIA within the combined sewer area range from $295 million to $1.7 billion. These strategies achieve 0.4 to 2.0 Bgal./year in overflow reductions.
- **I&I (Strategies 6 through 11):** Assuming $144 per linear foot and that 40 percent of all pipes in a target area will be repaired, I&I strategies cost $222 million or $343 million (depending on the scope of the target area). As expected, cost-effectiveness is better when focusing on a narrower area that contains the worst I&I offenders. For the screening strategies, we treat the costs as fixed (either $222 million or $343 million), and the level of I&I reduction achieved—either Low, Mid, or High (see Table 4.3)—is treated as an uncertainty. The cost-effectiveness of an I&I strategy will depend on the level of reduction achieved. For example, I&I High (8 percent) (Strategy 5) reduces roughly 1 Bgal. more in overflows than I&I Low (8 percent) (Strategy 1), although the strategy costs are assumed to be the same.

---

[3] The weighted average cost is $304,000, assuming 95 percent bioretention and 5 percent green roofs.

Figure 4.5
Screening Strategy Flow Reduction (2003 Typical Year) and Nominal Cost

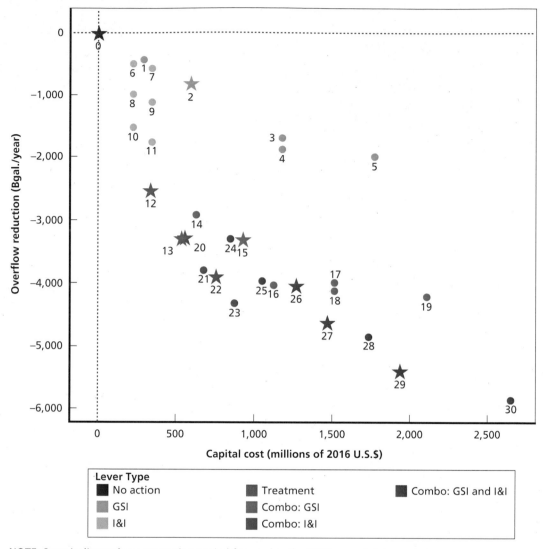

NOTE: Stars indicate those strategies carried forward in the RDM uncertainty analysis.
RAND RR1673-4.5

- **Treatment plant expansion (Strategy 12):** Assuming a nominal cost of $335 million and a reduction of 2.5 Bgal./year in overflows, the treatment plant expansion appears to be a very cost-effective option.
- **Treatment plant expansion and interceptor cleaning (Strategy 13):** SWMM modeling shows that cleaning the deep-tunnel interceptors provides roughly 760 Mgal./year in overflow reductions in addition to the effects from the treatment plant expansion. Assuming a nominal cost of $200 million to clean the deep interceptors ($535 million for the plant expansion, along with the intercep-

tor cleaning), this appears to be a cost-effective approach to reducing roughly 35 percent of systemwide overflows.

- **Combination strategies (Strategies 14 through 30):** Total capital costs for combination strategies range from roughly $560 million to $2.6 billion, with overflow reductions from 2.9 to 5.8 Bgal./year. The figure shows significant variations in cost-effectiveness. For example, Strategies 19 and 23 provide roughly the same overflow reduction (4.3 Bgal./year), but Strategy 23 is expected to cost nearly $1.3 billion less.

Conclusion

In this chapter, we formulated a set of regional, planning-level source reduction strategies for the ALCOSAN service area. Study Partners and Stakeholder Advisors helped identify policy levers for this analysis intended to either improve the function of the existing sewer system (e.g., expand the regional WWTP) or reduce the flow of stormwater into the system during rainfall events. We drew from these insights and the recent literature on stormwater management to identify four different policy lever types. We then used the SWMM simulation modeling framework to conduct a preliminary screening analysis of 30 different strategies, which include each of these levers in isolation or in combination. The screening process used a single set of assumptions regarding rainfall, land use, population, and cost and was intended to identify a promising subset to consider in the more-complete uncertainty analysis, which is presented in the following chapter.

All screening strategies were evaluated relative to an FWOA (i.e., no improvements to the existing infrastructure) and the 2003 Typical Year rainfall, which results in a baseline of 9.5 Bgal./year in total overflow. Under varying assumptions about the scale and effectiveness of the strategies, we find that the treatment plant alone reduced overflows by 25 percent and the treatment plant with cleaning the main interceptors reduced overflows by 34 percent; I&I strategies reduced total overflow 6 to 20 percent; and GSI strategies reduced overflows 5 to 21 percent. We tested a variety of combination strategies, which included a mix of the different lever types, finding overflow reductions from 35 to 62 percent. We estimate that the total capital costs of the strategies ranged from roughly $560 million to $2.6 billion.

We emphasize that this screening analysis was intended to provide preliminary insights and to help identify the subset of strategies for the final RDM analysis. There are significant uncertainties in the performance and cost of each of these strategies that are not yet captured in the results described in this chapter. As we will see in Chapter Five, cost and performance uncertainties are significant, and they should be at the forefront for decisionmakers when considering the path forward.

Robust Decision Making Strategy Comparison

The third and final phase of this pilot study was an initial quantitative scenario analysis using RDM methods. In this analysis, we tested a subset of the strategies introduced in Chapter Four against the future wastewater customer connection, land-use, and climate scenarios and incorporated cost and GSI performance uncertainties.

As noted in Chapter Four, none of the initial planning-level source reduction strategies considered in this analysis is able to eliminate SSO volumes, and most still yield substantial remaining CSO volumes and average hours in overflow per year. Incorporating future uncertainty only exacerbates this challenge, as shown later in this chapter, so the analysis in this report represents a first step rather than a comprehensive or conclusive solution. The goal in this chapter is to use RDM to help identify "low-regret" policy levers, or those that perform well in terms of benefits, costs, or cost-effectiveness across the full range of uncertainty considered. These low-regret levers could represent worthwhile near-term investments irrespective of final decisions made about other portions of the draft WWP or new infrastructure proposals that might emerge. This analysis is intended to inform an adaptive planning approach by highlighting potential low-regret, near-term options for ALCOSAN and the municipalities to consider.

We begin by estimating the overflow reduction performance of a selected subset of strategies across the range of overflow scenarios described in Chapter Three, along with a new scenario dimension representing uncertainty in the performance of GSI. Next, we introduce a set of additional uncertainties related to strategy costs. Results are presented for the selected strategies across the entire scenario ensemble in terms of total cost and cost-effectiveness, and we discuss which strategies are most cost-effective given different assumptions about future uncertainty. We also use "scenario discovery" methods to highlight which uncertain factors most often lead to poor cost-effectiveness for two strategy types, one focused on GSI and another combining a treatment plant expansion with source reduction. We conclude the chapter by highlighting several possible near-term, low-regret options for further consideration by regional planners.

The final strategy set and experimental design that support this chapter are described in Appendix E of this report.

Evaluating Strategies Against a Range of Overflow Scenarios

Strategies Considered

As discussed in Chapter Four, we identified a subset of strategies to test against a range of uncertain scenarios in the RDM analysis. This subset is summarized in Table 5.1, building on the detailed descriptions in Chapter Four and Appendix D. Note that every strategy except one includes an expansion of ALCOSAN's WWTP, a policy lever identified by study participants as an emerging and consensus next step and one for which ALCOSAN has already begun implementation.

Overflow Uncertainty

We modeled more than 500 unique scenario and year combinations throughout this analysis, which required more than 30,000 CPU-hours of computing time. This represents a significant leap forward in the application of high-performance computing for stormwater scenario analysis, but our ability to explore additional scenarios and strategies was nevertheless limited by the cost of these computing resources. As a result, in this portion of the analysis, we limited the number of years in the climate scenarios compared with the vulnerability analysis. Results in Chapter Three described a ten-year simulation period, which was then averaged to yield average annual overflow estimates. In this chapter, computing resources are instead balanced toward evaluating a higher number of combined strategies. We test strategies against 2003 Typical Year rainfall and the year 2013, which is a close match to the average annual rainfall statistics observed from 2004 through 2013 (Recent Historical). For the future climate

Table 5.1
Strategies Considered in Final RDM Analysis

ID	Strategy Name	Includes WWTP Upgrade	Includes Interceptor Clean	Includes GSI	Includes I&I Reduction
2	GSI-20			✓	
12	480 MGD	✓			
13	480 MGD + Clean	✓	✓		✓
15	480 MGD + GSI-20	✓		✓	
20	480 MGD + I&I Mid (8%)	✓			✓
22	480 MGD + Clean + I&I Mid (8%)	✓	✓		✓
26	480 MGD + GSI-20 + I&I Mid (8%)	✓		✓	✓
27	480 MGD + Clean + GSI-20 + I&I Mid (8%)	✓	✓	✓	✓
29	480 MGD + Clean + GSI-40 + I&I Mid (8%)	✓	✓	✓	✓

scenarios, we test strategies against the projected rainfall in 2047, which is the climate-adjusted version of 2013 and a similarly close match to the average the climate-adjusted rainfall from 2038 to 2047.

This represents a trade-off, because the single years selected for each climate scenario may not fully capture the range of outcomes or strategy performance. In addition, we are unable to consider the reliability of overflow reduction across different types of years. That said, the vulnerability analysis previously described provided useful insight into the selection of these single years, and using computing resources to test a wider range of strategies would be more useful at this stage of analysis.

GSI Performance Uncertainty

In this phase of analysis, we include a new scenario uncertainty dimension, GSI Performance, which varies a key performance parameter for GSI stormwater source reduction as identified in the GSI sensitivity testing (see Appendix D). In the Low performance scenario, we assume a 10:1 loading ratio. This implies that each acre of GSI captures runoff from 10 acres of DCIA, and the GSI is sized to capture runoff from 1 inch of rainfall (or 1.5 inches of rainfall in the GSI-40-AOP scenario). In the High performance scenario, alternatively, we assume GSI is strategically located in "high-flow" sites such that runoff from 25 acres of DCIA is routed to each acre of GSI (25:1 loading ratio). Note that, in the High GSI performance scenario, we are not adjusting the size of the GSI installations; we are adjusting only the runoff area that is tributary to the GSI. Given the substantial uncertainty associated with GSI in terms of flow reduction when implemented at a high number of sites across the region, these two scenarios were developed as bounding cases to help characterize the range of plausible performance. Strategies that include GSI are tested against both sets of assumptions in addition to the other scenario uncertainties. Table 5.2 shows the scenarios and years used in the final RDM analysis.

We considered a full factorial combination of the strategies in Table 5.1 and scenarios in Table 5.2 except for GSI infiltration, which was included only for the five strategies with GSI investments. Strategies without GSI were considered against 24 unique scenarios, while strategies with GSI were evaluated in 48 scenarios. This

Table 5.2
Overflow Scenarios Evaluated in Final RDM Analysis

Wastewater Customer Scenario	Land-Use Scenario	Climate Scenario	GSI Performance Scenario
Current Connections	Current Land Use	2003 Typical Year	Low: 10:1 loading ratio
Future Connections	SPC Growth	Recent Historical (2013)	High: 25:1 loading ratio
	2xPGH	Higher Intensity Rainfall (2047)	
		Higher Total Rainfall (2047)	

yielded $(4 \times 24) + (5 \times 48) = 336$ unique cases run as the combination of all strategies and scenarios.

GSI Results with Uncertainty

Using the experimental design described earlier, we first simulated the GSI-20 strategy—sized to control runoff from 20 percent of DCIA in the combined service area—across the range of overflow uncertainties considered (Table 5.3). The goal was to better understand how the benefits from large-scale GSI investment might vary with uncertainty before combining GSI with other policy levers. Table 5.3 divides out the land-use and climate scenario results in rows, and GSI performance scenarios as separate columns (Future Connections wastewater customer scenario only). The left-most column shows the total overflow results with no action. Under each GSI performance scenario, the total overflow (left) and change in overflow from no action (right) are shown, while the last column shows the difference in results from GSI performance uncertainty.

Assuming Current Land Use and 2003 Typical Year rainfall (first row), GSI-20 reduces total systemwide overflow 0.8 to 1.3 Bgal./year, with a difference of 0.5 Bgal./year between the Low and High GSI performance assumptions. However, GSI flow reduction increases in higher rainfall scenarios. Holding land use constant (Current Land Use), for example, GSI-20 overflow reduction increases between 0.9 to 1.3 Bgal./year in Recent Historical (2013 only), 1.6 to 2.0 Bgal./year in Higher Intensity Rainfall, and 1.6 to 2.1 Bgal./year in Higher Total Rainfall. The GSI Performance scenario difference between Low and High assumptions remains relatively constant at 0.4 to 0.5 Bgal./year with Higher Intensity Rainfall.

By contrast, the performance of GSI-20 remains constant or declines slightly across land-use scenarios with greater impervious cover. For example, the overflow reduction when holding rainfall constant at the 2003 Typical Year scenario is very similar for Current Land Use (0.8 to 1.3 Bgal./year), SPC Growth (0.8 to 1.3 Bgal./year), and 2xPGH (0.7 to 1.2 Bgal./year), respectively (first, fifth, and ninth rows of Table 5.3).

The GSI-20 scenario results yield several insights. The first is that, under current rainfall and land-use assumptions, overflow reduction from installing GSI sized for 20 percent of impervious cover across the combined service area is modest and addresses only 8 to 13 percent of the total sewer overflow challenge. However, GSI does yield greater flow reduction with higher future rainfall. In the future climate scenarios considered, GSI-20 performance doubles despite more modest increases in total overflow, so that GSI is able to address upward of 13 to 16 percent of total overflow under these assumptions. This difference in performance has a notable effect on the cost-effectiveness results discussed later in this chapter.

Table 5.3
Overflow Simulation Results from GSI-20 with Uncertainty (Bgal./year)

Land-Use Scenario	Climate Scenario	No Action	GSI Performance Scenario				GSI Scenario Change
			Low		High		
			GSI-20	Overflow Reduction	GSI-20	Overflow Reduction	
Current Land Use	2003 Typical Year	10.0	9.2	0.8	8.7	1.3	0.5
Current Land Use	Recent Historical (2013)	11.6	10.7	0.9	10.3	1.3	0.4
Current Land Use	Higher Intensity Rainfall (2038)	12.3	10.8	1.6	10.4	2.0	0.4
Current Land Use	Higher Total Rainfall (2038)	12.9	11.2	1.6	10.8	2.1	0.4
SPC Growth	2003 Typical Year	10.4	9.5	0.8	9.1	1.3	0.4
SPC Growth	Recent Historical (2013)	12.1	11.1	1.0	10.7	1.4	0.4
SPC Growth	Higher Intensity Rainfall (2038)	12.9	11.2	1.7	10.8	2.1	0.4
SPC Growth	Higher Total Rainfall (2038)	13.3	11.7	1.6	11.2	2.1	0.4
2xPGH	2003 Typical Year	11.0	10.2	0.7	9.8	1.2	0.5
2xPGH	Recent Historical (2013)	12.7	11.9	0.8	11.5	1.3	0.4
2xPGH	Higher Intensity Rainfall (2038)	13.6	12.0	1.6	11.6	2.0	0.4
2xPGH	Higher Total Rainfall (2038)	14.0	12.5	1.5	12.1	1.9	0.4

NOTE: Table shows Future Connections wastewater customer scenario results only. All overflow values would be reduced by 0.5 Bgal./year in the Current Connections scenario.

Overflow Reduction Results from All Strategies with Uncertainty

Overflow results for the selected strategies across all uncertain scenarios are summarized in Figure 5.1. Each box plot shows the range of total overflow (Bgal./year) simulated in the SWMM model with that strategy implemented across all uncertain futures simulated, summarizing either 24 or 48 futures as noted earlier. Note that we use box plots as a convenient means of summarizing the range of uncertainty, but this is not to imply that any given point in the distribution is more or less likely than another. In other words, we make no assumption about the likelihood of these scenarios, but rather seek to identify the range of plausible results to better characterize future vulnerability.

Figure 5.1
Remaining Overflow with Selected Strategies, All Scenarios

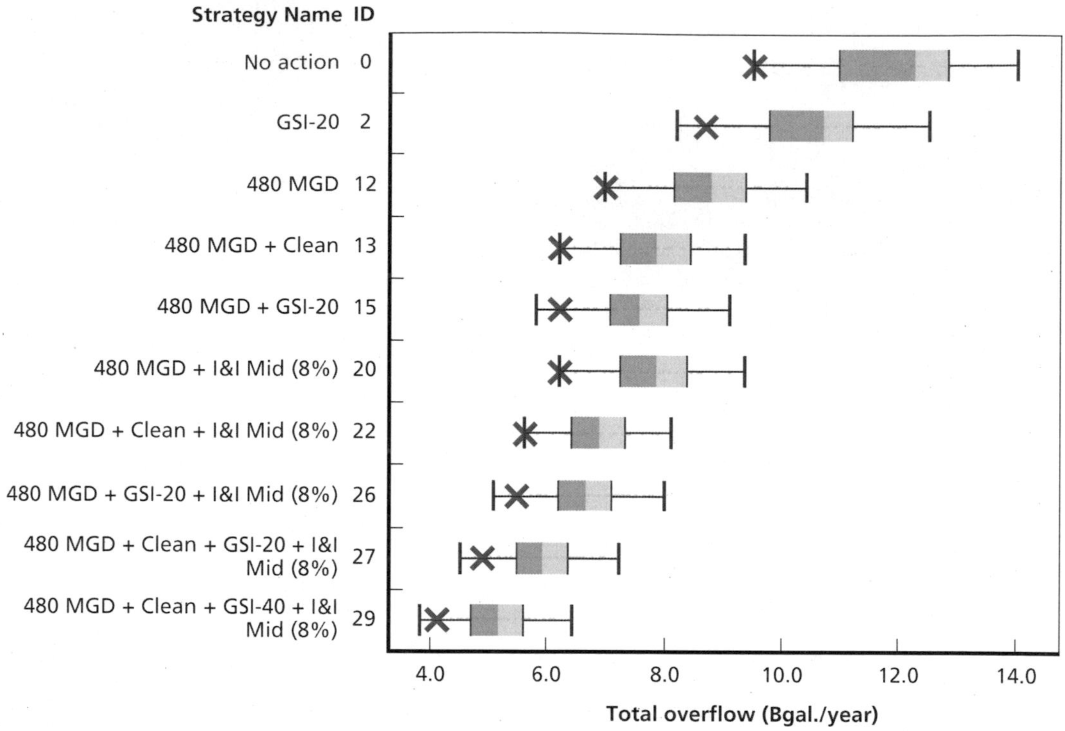

NOTE: The box plots presented do not represent probability distributions but instead report the results of a set of model runs (scenarios). Strategies with GSI are evaluated in 48 unique scenarios, and those without GSI show 24 scenario results. Each point summarized represents one mapping of assumptions to consequence, and the points are not assumed to be equally likely. Components of the box plot include the 25th and 75th percentiles (edges of each box), median (vertical line where the two gray shaded areas meet), and extremes of the data set (whiskers). The blue X indicates an initial set of assumptions used in strategy screening that are similar to those applied in other recent research. 480 MGD = treatment plant expansion; Clean = deep-tunnel interceptor cleaning.
RAND RR1673-5.1

In this and subsequent plots, the results simulated with the nominal assumptions used in the Chapter Four screening analysis—2003 Typical Year hydrology, Current Connections, Current Land Use, and Low GSI Performance—are identified with an "X."

The first row in the figure shows the range of results in an FWOA (no action), mirroring the detailed investigation of future vulnerability described in Chapter Three.[1] Total overflow without action range from 9.5 to 14 Bgal./year across all scenarios. The second row summarizes the results with the GSI-20 (Strategy 2) results described earlier, yielding a range of 8.2 to 12.5 Bgal./year.

---

[1]   Overflow results differ slightly from those in Chapter Three because the final analysis uses the single years 2003, 2013, and 2047 to represent climate uncertainty rather than a ten-year average.

The selected strategies simulated improve on the FWOA and GSI-20 results. For instance, upgrading the treatment plant (Strategy 12) reduces total overflow to 7.0 to 10.4 Bgal./year, a reduction of 2.5 to 3.7 Bgal./year. Adding system improvement or source reduction policy levers further reduces overflow relative to the FWOA. Strategy 13, which includes the treatment plant expansion and cleaning the main interceptors, reduces overflows to 6.2 to 9.4 Bgal./year (a reduction of 3.3 to 4.7 Bgal./year) across the uncertain scenarios, a further reduction of 800 to 900 Mgal./year compared with Strategy 12. Coupling treatment plant expansion with the GSI-20 lever sized to control runoff from 20 percent of impervious area in the combined sewer area (Strategy 15) similarly reduces overflow by 3.3 to 5.3 Bgal./year but leaves 5.8 to 9.1 Bgal./year of overflow remaining.[2] Adding a pipe repair program to reduce inflows and infiltration (480 MGD + I&I Mid [8%], Strategy 20) to the treatment expansion also shows a similar range of results.

The final four strategies considered in the uncertainty analysis include a combination of three or more levers: treatment expansion, interceptor cleaning, I&I, and/or GSI. These strategies unsurprisingly yield the most overflow reduction, but the benefits generally show some diminishing returns from multiple levers together as discussed in Chapter Four.

Specifically, combining treatment expansion, a main interceptor clean, and a mid-range pipe repair program (Strategy 22) reduces total overflow by 3.9 to 5.9 Bgal./year (5.6 to 8.1 Bgal./year remaining). Similarly, Strategy 26 combines treatment, GSI-20, and pipe repair and yields a reduction of 4.0 to 6.2 Bgal./year (5.1 to 7.6 Bgal./year of total overflow remaining). Adding all four levers together using GSI-20 yields further improvement (Strategy 27), while substituting in GSI-40 (GSI designed to control 40 percent of the impervious area; Strategy 29) yields the greatest total overflow reduction among the strategies considered (5.3 to 8.0 Bgal./year). Note, however, the substantial overlap in overflow reduction for many of the strategies compared across the uncertainty range. With any given set of scenario assumptions, the ranking might be evident, but, in general, these results show that many different strategies could produce similar volumes of reduced overflow.

These results show that source reduction, combined with treatment expansion and operational improvements, can yield significant overflow reduction. Further, these benefits appear to increase in scenarios in which total overflow is higher. However, the range of simulation results also shows the limits of the strategies evaluated in this pilot effort. For one, the nominal assumptions used in the screening analysis nearly always

---

[2]   Strategy 15 is similar to a strategy tested by PWSA that also includes a 480-MGD plant expansion, along with management of 1,835 impervious acres (PWSA, 2016). PWSA showed total overflow reduction of 4.09 Bgal./year with this strategy, while comparable results from this analysis (2003 Typical Year, Current Land Use, Current Customer Connections) yielded a systemwide reduction of 3.3 Bgal./year. Results are generally similar, but the more targeted and refined approach in PWSA analysis, focusing on priority sewersheds and developing more detailed design concepts, likely explains the difference in performance.

yield the most optimistic or near-best results in terms of meeting the regulatory target of eliminating or greatly reducing overflows. In other words, across a range of assumptions about future climate, customer connections, and land use, the overflow results generally increase across all strategies, making the target harder to achieve.

Moreover, despite testing a wide range of aggressive source reduction policy levers, including those that may not be feasible when incorporating a more realistic range of constraints (e.g., GSI-40), none of the strategies considered reduces overflows close to 0. Strategy 29, an "all of the above" strategy that includes significant stormwater source reduction through pipe repair and GSI, can yield 3.8 Bgal./year of remaining overflows in the most optimistic set of assumptions but nevertheless still yields nearly 6.5 Bgal./year at the high end of the range.

This is also true specifically for SSO reduction, as discussed in Chapter Four. None of the strategies considered leads to elimination or near elimination of SSOs in any of the scenarios considered. Strategies with three or more levers included reduce SSO volumes by approximately 250 to 470 Mgal./year but leave 300 to 600 Mgal./year remaining (not shown; see Figure 4.4 in Chapter Four for screening results).

As a result, no strategy considered in this initial analysis can reliably meet the region's planning or regulatory goals without including additional infrastructure investments, such as additional conveyance or deep-tunnel storage, as identified in ALCOSAN's draft WWP.

Additional Performance and Cost Uncertainties

Next, we consider strategy performance in terms of cost and cost-effectiveness. This analysis builds on the preliminary capital cost estimates introduced in Chapter Four, with the same important caveat that these represent first-order estimates of capital cost only and do not account for the life-cycle costs (capital, O&M, and financing) or implementation time associated with the strategies.

Incorporating preliminary capital costs entailed expanding the range of uncertain factors to include uncertainties related to strategy cost (see Chapter Three, Table 3.1) and developing quantitative experiments with both overflow and cost uncertainty. Specifically, we developed a separate sampling design for the cost uncertainty factors, which were then combined with the overflow scenarios to generate the final ensemble of cases.

Capital Cost Uncertain Factors

With guidance from the Study Partners and Stakeholder Advisors, we identified one or more key uncertainties related to capital cost for each type of policy lever. These uncertainties are briefly summarized in the following section, and the uncertain factors and ranges chosen are discussed in detail in Appendix D of this report. Table 5.3 provides the final ranges for all uncertain cost factors.

GSI Cost Uncertainty

The literature shows a wide variation of potential GSI capital costs, both based on observed costs from completed projects and based on engineering, or "bottom-up" cost estimates (ALCOSAN, 2015a; District of Columbia Water and Sewer Authority, 2015; O'Donnell et al., 2014; Philadelphia Water Department, 2009; Valderrama et al., 2013; Water Environment Federation, 2015). We represent this uncertainty through three uncertain factors: (1) *bioretention cost per impervious acre controlled*, which represents the primary GSI type simulated in this study, as well as similar retention, and/or infiltration-based approaches; (2) *green roof cost per acre controlled*, which is represented separately because of the higher per-acre costs using green roofs; and (3) *percentage green roof*, which represents the proportion of a GSI strategy that would necessitate green roofs and scales costs upward as an increased percentage is needed. As shown in Table 5.4, a wide range of plausible assumptions was included for all three factors given current best practice and knowledge about implementing GSI at a large scale.

I&I Reduction Cost Uncertainty

Reducing I&I requires repairing pipes and manholes to reduce both RDII and GWI. The total strategy cost is a function of (1) the assumed cost per manhole, (2) the cost per linear foot to repair pipes (*pipe repair cost*), and (3) the assumed percentage of pipes and manholes in an area that need to be repaired to achieve the target I&I reduction

Table 5.4
Final Cost Uncertainty Factors and Ranges for RDM Analysis (2016 Dollars)

Cost Uncertainty	Low	Nominal	High
Bioretention cost ($ per acre controlled)	154,000	285,000	554,000
Green roof cost ($ per acre controlled)	571,000	672,000	772,000
Percentage green roof (%)	0	5	15
Pipe repair cost ($ per linear foot)	86	144	222
Percentage of pipes and manholes repaired (%)	20	40	100
Treatment plant expansion cost (millions of $)	234	335	502.5
Deep-tunnel interceptor cleaning cost (millions of $)	140	200	300

NOTE: GSI costs (bioretention and green roof) were sampled jointly rather than independently, reducing the number of sampling dimensions by one. Costs reflect a conversion to 2016 dollars for a common baseline using a consumer price index inflator, and therefore differ slightly from those listed in Appendix D tables that draw from the original sources.

(*percentage of pipes and manholes repaired*). The latter two parameters are included as cost uncertainties in this analysis.

The pipe repair costs are based on three data sources; the low and mid-values are based on cost data from actual repair efforts in the region (ALCOSAN, 2015b), and the high value is the default for 8-inch pipes in the Philadelphia Water Department (Philadelphia Water Department, 2009). The percentage of pipes or manholes repaired ranges from 20 to 100. At the low end, this would assume substantial economies of scale from flow-monitoring studies and other system monitoring, which could be used to target relatively small sections of pipe that contribute the majority of I&I for a more cost-effective repair strategy. The high end, alternatively, assumes that 100 percent of pipes and manholes in a target area would need to be repaired to achieve the targeted reduction level. This is consistent with the assumptions made by ALCOSAN in its recent source reduction study (ALCOSAN, 2015b).

Treatment Plant Expansion Cost Uncertainty

Finally, we adapted the capital cost assumptions from ALCOSAN's draft WWP for treatment plant expansion to 480 MGD. Following ALCOSAN's planning cost assumptions, this analysis assumes that capital costs could vary between –30 percent (low) and +50 percent (high) of the initial capital cost estimate.[3] This cost range represents such uncertainties as household wastewater rates and bond interest rates, all of which will affect the final cost of the treatment plant expansion (ALCOSAN, 2012f, p. 11-58).

Deep-Tunnel Interceptor Cleaning Uncertainty

Based on preliminary PWSA planning, the cost for cleaning the deep-tunnel interceptors is estimated at approximately $200 million (PWSA, 2016). This amount factors in the new drop shafts that would need to be created to get cleaning machinery down into the tunnels, along with the cost of the cleaning itself. A nominal value of $200 million is used in this analysis when examining the cost and cost-effectiveness of cleaning the existing deep-tunnel interceptors. As with the treatment expansion, this analysis assumes that capital costs could vary between –30 percent (low) and +50 percent (high) of the initial capital cost estimate.

Final Ensemble of Scenarios

Using these cost uncertainty factors and ranges, we used a Latin hypercube sampling (LHS) approach to develop an efficient sample of all dimensions.[4] LHS is a "space-filling" statistical method that helps ensure that the entire distribution of each param-

---

[3]   Note that all values are converted to 2016 dollars using a consumer price index inflator for ease of comparison.

[4]   The two GSI cost per acre controlled parameters (bioretention and green roof costs) were sampled jointly rather than independently to reduce the number of dimensions and yield a single GSI cost per acre uncertainty range.

eter is sampled consistently. The approach is often used to develop efficient samples in support of RDM analysis and scenario discovery (Bryant and Lempert, 2010; Groves and Lempert, 2007; Lempert, Popper, and Bankes, 2003). We developed a 100-point LHS sample across the cost uncertainty dimensions listed in Table 5.3 and included several additional samples representing either the extreme end points of the distribution (all values set to their lowest or highest values) or the nominal cost estimates described in Chapter Four. This yielded a total sample of 103 points, which was then combined with the 48 overflow scenario and year combinations earlier in this chapter (48 × 103) to produce a final ensemble of 4,944 uncertain scenarios. The cost and cost-effectiveness results discussed through the remainder of this chapter describe strategy performance across this ensemble of roughly 5,000 scenarios.

Strategy Costs with Uncertainty

Cost uncertainty ranges using this approach are summarized in Figure 5.2 for the final set of strategies. The box plots summarize the sample of 103 cost uncertainty scenarios, and an "X" once again indicates the nominal cost assumption used for initial screen-

Figure 5.2
Range of Selected Strategy Costs, All Scenarios

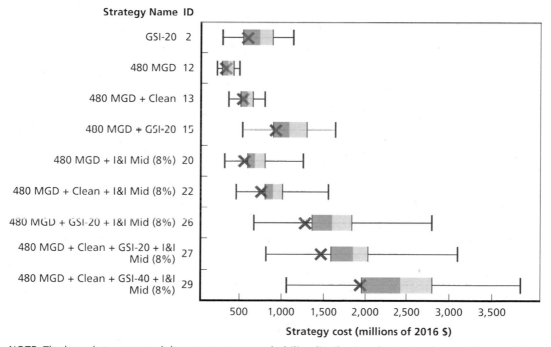

NOTE: The box plots presented do not represent probability distributions but instead report the results of a set of model runs (scenarios). Each point summarized represents one mapping of assumptions to consequence, with 103 scenarios in total for each strategy, and the points are not assumed to be equally likely. The blue X indicates the nominal assumptions used in the screening analysis.
RAND RR1673-5.2

ing in Chapter Four. These results show that the range of cost uncertainty generally scales with the level of effort or investment. For example, the cost of upgrading the treatment plant alone (Strategy 12) ranges from $235 million to $503 million (2016 dollars). Adding interceptor cleaning (Strategy 13) expands the range to $375 million to $803 million; alternatively, adding GSI-20 to the treatment expansion (Strategy 15) greatly expands the range to between $534 million and $1.6 billion. Including various combinations of treatment plant expansion, interceptor cleaning, GSI, and I&I reduction together further increases the level of uncertainty; Strategy 26, for instance, yields costs ranging from $676 million to $2.8 billion, and Strategy 29 includes a wider cost range.

These results confirm and reinforce the level of uncertainty currently associated with large-scale source reduction for a regional sewer system like ALCOSAN's. Over a long period of implementation (ten to 20 years), costs could plausibly end up at the lower end because of technological improvement, evolving best practice, and economies of scale. Alternatively, higher costs also remain plausible as source reduction is scaled up, recognizing that additional locations identified for GSI or I&I reduction might end up incrementally more expensive to address as sites with the lowest barriers to implementation are already addressed.

Strategy Cost-Effectiveness Results

We next turn to a consideration of strategy cost-effectiveness, combining the uncertainty results for both overflow reduction and strategy cost. For these results, we consider strategy performance across all of the 4,944 uncertain scenarios. Figure 5.3 shows a box plot summary of cost-effectiveness results, represented in terms of 2016 dollars per gallon of total overflow reduced, for eight strategies. Each box plot summarizes strategy performance across all overflow and capital cost uncertainties (4,944 scenarios). A lower value in this metric is better, implying a lower cost to achieve each additional gallon of annual overflow reduction. An "X" denotes the scenario assumptions used in the screening analysis. A line is included at $0.35 per gallon, which is the approximate cost-effectiveness estimated by ALCOSAN for the draft WWP using its nominal cost and performance assumptions.[5] While we use this as a convenient reference point, the strategies explored here are not necessarily directly comparable to ALCOSAN's plan. A cost-effectiveness result that varies from this reference point may be due to a higher or lower scale of intervention or a different balance between CSO and SSO reduction. Nonetheless, we use $0.35 per gallon as a useful threshold to differentiate between acceptable and poor cost-effectiveness.

---

[5]   Ongoing negotiations between USEPA, PADEP, and regional stakeholders may result in a new plan with different cost and performance estimates.

Figure 5.3
Cost-Effectiveness for Selected Strategies, All Scenarios

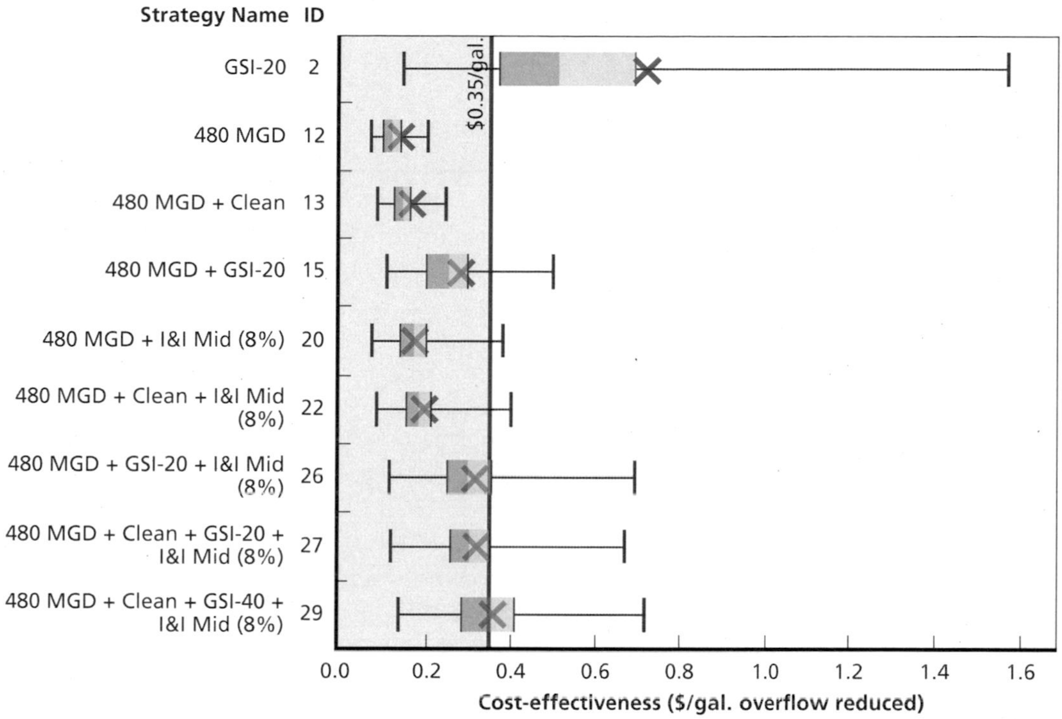

NOTE: The box plots presented do not represent probability distributions but instead report the results of a set of model runs (scenarios). Strategies with GSI are evaluated in 4,944 unique scenarios, and those without GSI show 2,472 scenario results. Each point summarized represents one mapping of assumptions to consequence, and the points are not assumed to be equally likely. Components of the box plot include the 25th and 75th percentiles (edges of each box), median (vertical line where the two gray areas meet), and extremes of the data set (whiskers). The blue X indicates the set of assumptions used in the screening analysis. Cost is represented in 2016 constant dollars, and the red reference line shows the average cost-effectiveness of ALCOSAN's draft WWP. Yellow shading indicates cost-effectiveness below this threshold.

RAND RR1673-5.3

The first row shows the cost-effectiveness results for GSI-20 alone. The range is very wide depending on the overflow and cost scenario assumption, ranging from $0.14 to $1.59 per gallon, with many scenarios above the reference threshold of $0.35 per gallon. We investigate these results further using scenario discovery methods, described below.

All other strategies shown combine treatment expansion with one or more other levers. the range of cost-effectiveness results for these strategies scales with the level of intervention, which is similar to the cost results. Unsurprisingly, the narrowest range is associated with the treatment plant expansion, which shows good performance regardless of the cost assumption made ($0.07 to $0.20 per gallon). In other words, across all scenarios, the overflow reduction achieved by a treatment plant expansion is always or nearly always highly cost-effective, even in scenarios in which expansion costs are up to 50 percent more expensive than in initial WWP estimates. Including interceptor cleaning with treatment plant expansion (Strategy 13) yields similarly acceptable but somewhat higher cost-effectiveness results across the full range of scenario assumptions considered.

Next, including I&I reduction (Strategy 20) or both interceptor cleaning and I&I reduction (Strategy 22) increases the cost-effectiveness range, particularly at the higher end, leading to upper bounds of $0.38 per gallon or $0.39 per gallon, respectively. At the upper end, the latter values are slightly above the reference value from the draft WWP, while, at the lower end of the range, they show very good cost-effectiveness performance.

Given that both GSI costs and GSI performance are very uncertain, the ranges are generally greater for strategies combining treatment expansion and GSI. For example, the cost-effectiveness of Strategy 15 (480 MGD + GSI-20) ranges from $0.10 to $0.50 per gallon (depending on the scenario assumption). At the low end, this would be a highly cost-effective strategy; at the high end, alternatively, an inefficient outcome. These ranges continue to expand upward as the level of investment scales up. Strategy 26 (480 MGD + GSI-20 + I&I Mid [8%]) shows cost-effectiveness results ranging from $0.10 to $0.70 per gallon. Strategies 27 and 29 also include interceptor cleaning and yield similarly very wide cost-effectiveness ranges.

Strategy 27, to which we refer as the *Combined Source Reduction Strategy* through the remainder of this discussion for convenience, includes a treatment plant expansion, interceptor cleaning, GSI-20 (sized to control runoff from 20 percent of impervious area), and pipe repair intended to reduce 20 to 40 percent of current RDII flow for targeted high-flow areas. The scale of GSI investment in this strategy is substantial but below the targets currently set recently by such cities as Philadelphia or Milwaukee (see Chapter Four).

The strategy yields notable overflow reduction (Figure 5.1), with cost-effectiveness ranging from $0.11 per gallon to $0.67 per gallon, but includes a number of plausible of scenarios with excessively high per-gallon costs. Specifically, in approximately 25 percent of scenarios, its cost-effectiveness is worse than the average value achieved in the draft WWP. Note that similar strategies, including combinations of treatment plant expansion, GSI-10 or GSI-20, and I&I Low or I&I High, which are not tested in this analysis, would likely achieve a similar balance between overflow reduction and cost-effectiveness.

Identifying Lower-Regret Policy Levers

Comparing Strategy "Regret"

The results shown above summarize and compare strategy performance across the full range of uncertain scenarios. Another way to compare strategies, however, is to consider their performance or ranking in each specific scenario and then estimate how much worse a given strategy performs than the best-performing strategy in that scenario. This builds on the concept of "regret" (Lempert, Popper, and Bankes, 2003; Savage, 1954), essentially quantifying how much a decisionmaker would value the best strategy in a given realization of the future when compared with the actual strategy chosen. The goal or criterion might be to select a strategy that minimizes average or median regret across all scenarios or, alternatively, that minimizes regret in the worst scenario for that strategy.

One limitation of this approach is that it allows for comparisons only between strategies actually evaluated, so that regret in each scenario is calculated relative to the best strategy considered, not necessarily the best possible strategy. Nevertheless, we used the ensemble of scenarios described above to estimate regret in terms of overflow reduction (Bgal./year reduced) or cost-effectiveness (cost per gallon) for eight strategies. Strategy rankings using this metric when looking across all scenarios for each metric are similar to those already shown in Figures 5.1 and 5.3 and are omitted here. But a summary statistic (median regret) can help to show trade-offs between overflow reduction and cost-effectiveness.

Figure 5.4 provides an illustration. The figure includes regret in terms of overflow reduction (y-axis) and cost-effectiveness (x-axis) for each of the eight remaining strategies. Points are sized to show the median remaining overflows with the strategy in place. An ideal strategy would yield low regret across both dimensions (results close to the origin). For example, Strategy 12 yields median overflow reduction regret of 3.5 Bgal./year, meaning that the highest investment strategies yield much higher overflow reduction than this strategy at the median and the regret is high. However, median cost-effectiveness regret for this strategy is 0, suggesting that the strategy is nearly always one of the best or near-best performers in terms of cost-effectiveness. The reverse is true for Strategy 29, which yields low median overflow reduction regret but the highest median cost-effectiveness regret.

In this way, Figure 5.4 traces out a robustness trade-off curve between overflow reduction and cost-effectiveness and can help ALCOSAN and municipal planners identify strategies that might better balance both goals and achieve acceptable performance when they face future uncertainty.

The regret comparison does not show any strategies close to the origin that balance both overflow reduction and cost-effectiveness regret. At the median, including interceptor cleaning reduces overflow regret without substantial increases in cost-effectiveness regret (compare Strategies 26 and 27, for example). Otherwise, the trade-off appears primarily linear; adding levers or higher increments of investment reduces overflow regret but generally increases cost-effectiveness regret.

The Combined Source Reduction Strategy (Strategy 27) is on the lower boundary of the curve and represents a potentially promising combined approach, yielding median overflow regret below 1 Bgal./year with median cost-effectiveness regret

Figure 5.4
Overflow and Cost-Effectiveness Median Regret, Selected Strategies

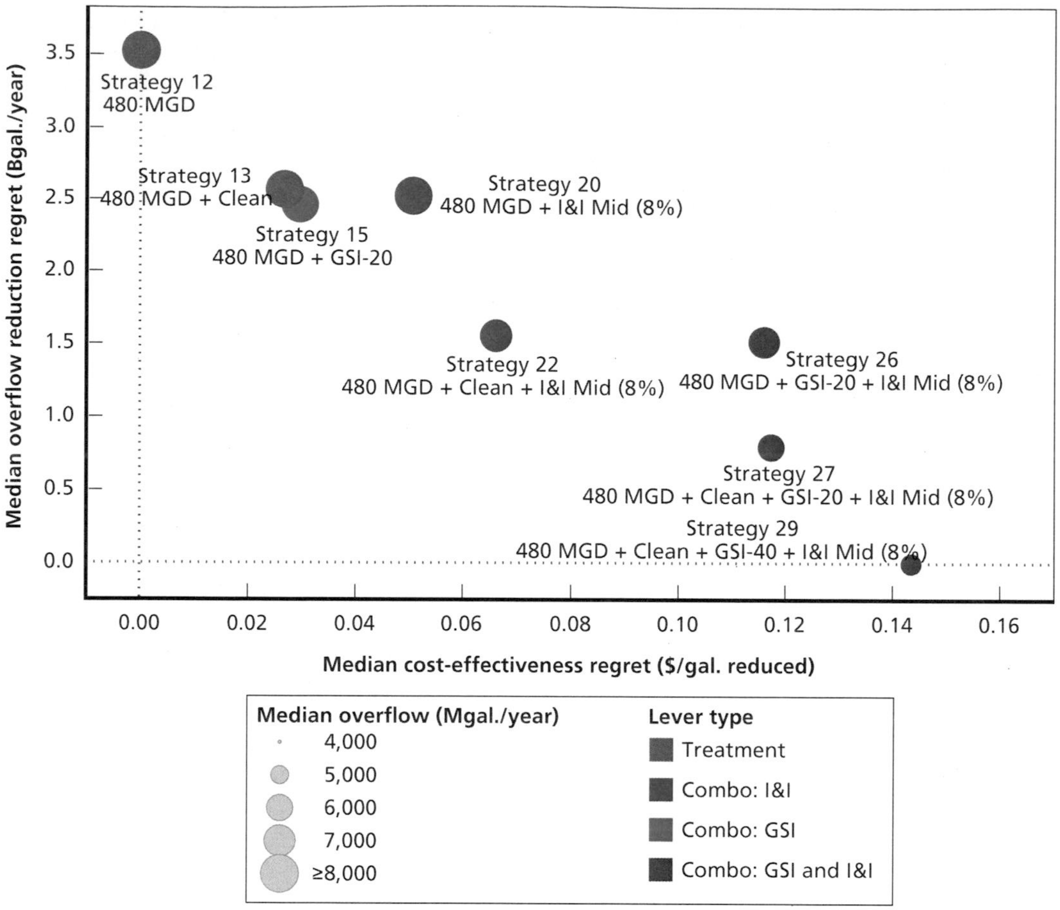

NOTE: Strategy 2 (GSI-20) results would appear above the upper-right corner of the plot area and are omitted for clarity.
RAND *RR1673-5.4*

at approximately $0.12 per gallon. We focus on the Combined Source Reduction Strategy, along with a strategy including only GSI (GSI-20; Strategy 2), through the remainder of this analysis to help illustrate how future uncertainty is leading to good or poor performance in terms of cost-effectiveness.

Scenario Discovery to Identify Key Uncertain Drivers

This analysis considers uncertainty from a wide range of sources. The results from the Combined Source Reduction Strategy, for example, vary across all nine uncertain factors described above. However, not all of these uncertain factors contribute equally to strategy performance. As a result, we next use scenario discovery to help identify those drivers that most commonly lead to poor or unacceptable cost-effectiveness performance for either the Combined Source Reduction Strategy or GSI-20 alone (Bryant and Lempert, 2010; Groves and Lempert, 2007).

As a starting point, Figure 5.5 illustrates how cost-effectiveness varies with remaining overflows (y-axis) and capital cost (x-axis) for the Combined Source Reduction Strategy. Each point in this plot represents one of the nearly 5,000 scenarios considered, and the colors illustrate the resulting cost-effectiveness. Notably, the cost-effectiveness performance is poor in cases in which (1) capital costs are at the high end, as expected, and (2) remaining overflow is at the low end of the uncertainty range. In other words, this strategy is more cost-effective in higher-flow than lower-flow futures, even with higher-than-expected capital costs.

GSI-20 Results

As noted in Figure 5.3, the range of cost-effectiveness results is especially broad for strategies that include GSI, given the uncertainty surrounding both overflow reduction and cost. Figure 5.5 further investigates this uncertainty for Strategy 2, which includes GSI only. Scenario discovery methods were used to identify the uncertain drivers that yield cost-effectiveness lower than $0.35 per gallon—focusing on the acceptable rather than poor or unacceptable scenarios given the large number of scenarios yielding poor cost-effectiveness when implementing GSI-20 alone.

For Strategy 2, 22 percent of the 4,944 scenarios considered yielded cost-effectiveness results lower than $0.35 per gallon. The results of a scenario discovery investigation suggested that three uncertainties—average annual rainfall, average GSI cost per acre, and GSI performance—can be used to distinguish acceptable from unacceptable performance.[6] Figure 5.5 shows the cost-effectiveness from all scenario results for Strategy 2, with each point in the plot representing one scenario. Gray points represent scenarios with cost-effectiveness higher than $0.35 per gallon, while green points represent those within the acceptable range lower than $0.35 per gallon. The climate and GSI performance scenarios are divided into separate rows, and the x-axis represents the assumption for average GSI capital cost per impervious acre controlled. The areas highlighted in yellow show the set of assumptions that that yield acceptable cost-effectiveness, with the dashed red lines showing the specific thresholds identified in part with statistical tools.[7]

---

[6]   Average GSI cost per acre is calculated as the weighted average of per-acre bioretention costs and green roof costs. A separate scenario assumption regarding the percentage of green roofs included (ranging from 0 to 15 percent) (Table 5.3) is used for the proportional weight in each scenario. Therefore, this calculation combines three scenario uncertainties into a single input used for scenario discovery. Given these assumptions, average GSI cost per acre is dominated by bioretention costs, with a small increase in the upper bound as additional green roofs are incorporated.

[7]   This analysis used the Patient Rule Induction Method (PRIM) as implemented in the R statistical software package "sdtoolkit" (Bryant and Lempert, 2010). Using the algorithm, we identified three separate PRIM "boxes" that correspond to historical rainfall (High GSI performance), future rainfall (Low GSI performance), and future rainfall (High GSI performance). GSI cost thresholds identified for each of these cases varied as shown

Figure 5.5
Overflow and Cost Results for a Combined Source Reduction Strategy (Strategy 27) Across All Scenarios

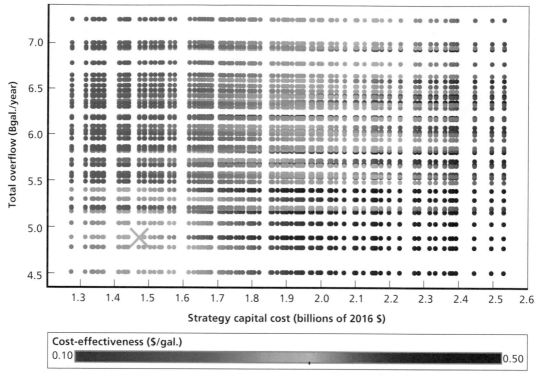

NOTE: Plot summarizes results from 4,944 unique scenarios. Each point summarized represents one mapping of assumptions to consequence, and the points are not assumed to be equally likely. The X indicates the nominal assumptions used in the screening analysis for reference. Red shades indicate scenarios in which this strategy yields excessively high costs on a per-gallon basis (higher than $0.35 per gallon).
RAND *RR1673-5.5*

The first row of Figure 5.6 shows that GSI-20 cost-effectiveness is always or nearly always higher than $0.35 per gallon when assuming Low GSI performance and either 2003 Typical Year or Recent Historical rainfall, respectively. With the High GSI performance assumption (second row), alternatively, a low number of scenarios yield values lower than $0.35 per gallon, when also assuming that average GSI per-acre costs are less than $230,000. In other words, under existing rainfall conditions, one has to carefully manage GSI costs and identify high-flow sites (25:1 loading ratio) to achieve cost-effective performance.

in Figure 5.6. These three boxes together capture 94 percent of the scenarios with cost-effectiveness lower than $0.35 per gallon (coverage), and 97 percent of scenarios with these characteristics yield below-threshold cost-effectiveness (density).

Figure 5.6
Scenario Discovery Results for GSI-20 (Strategy 2)

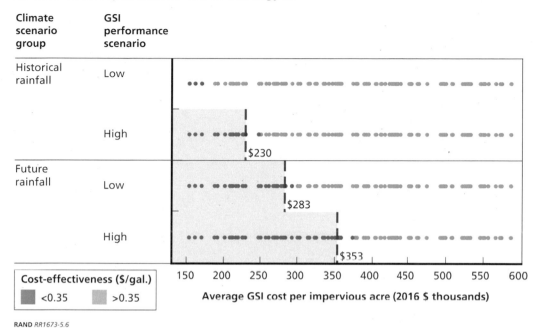

RAND *RR1673-5.6*

This pattern changes when we look at the climate scenarios with plausible future rainfall increases (bottom half of plot). Here, there is still variation with GSI performance, but the assumption of increasing future rainfall leads to many more scenarios with acceptable cost-effectiveness performance when coupled with lower GSI installation costs. Specifically, scenario discovery showed that future rainfall combined with either an average GSI cost per impervious acre below $283,000 and High GSI performance, or below $353,000 with Low GSI performance, very often leads to GSI cost-effectiveness below $0.35 per gallon.

In general, these results suggest two key points. First, assumed GSI performance unsurprisingly influences cost-effectiveness, with modest but clear improvements in the High GSI scenario, in which we assume that GSI is located in strategic, high-flow sites and is able to capture runoff from a much larger portion of each subcatchment. Second, assumptions about future rainfall appear to be important drivers. Under historical rainfall assumptions and standard GSI performance assumptions, GSI does not appear to be cost-effective when considering its sewer overflow benefits alone. Even across a range of assumptions regarding GSI cost, the GSI-20 strategy performs poorly relative to other strategies targeting sewer overflow reduction in this analysis using either 2003 or 2013 hydrology. However, GSI becomes increasingly cost-effective as average annual rainfall increases and the climate scenario becomes more adverse. In these cases, each acre of GSI is storing higher rainfall volumes and yielding greater benefit, which leads to improved cost-effectiveness. This suggests that taking into account future uncertainty could make GSI investments more appealing to municipalities in the ALCOSAN service area when compared with current planning assumptions alone.

Combined Source Reduction Strategy Results

We also used scenario discovery statistical algorithms to help investigate what drives the Combined Source Reduction Strategy to perform well or poorly in terms of cost-effectiveness. For this combined strategy, more scenarios yielded acceptable cost-effectiveness performance, so the approach was reversed to identify assumptions leading to *poor* rather than *acceptable* cost-effectiveness outcomes—that is, identify which uncertain factors were most often leading to cost-effectiveness results in excess of $0.35 per gallon (Figure 5.5, red-shaded region). For this strategy, the results showed that a similar subset of uncertainties—in this case, average annual rainfall, average GSI cost per acre, and proportion of pipe to repair to achieve the specified level I&I flow reduction—describes nearly all of the poor cost-effectiveness performances.

The results are summarized in Figure 5.7.[8] This figure again shows a scatterplot for the Combined Source Reduction Strategy, with each point representing one scenario realization. Each pane shows results in two of the four rainfall scenarios (2003 Typical Year and Recent Historical, respectively). The y-axis is the average cost per controlled acre for GSI, while the x-axis shows the pipe repair assumption ranging from 20 to 100 percent. Any scenario with cost-effectiveness higher than $0.35 per gallon is highlighted as a solid point in red, while those with lower than $0.35 per gallon performance are shown as open gray circles.

Cost-effectiveness is always or nearly always lower than $0.35 per gallon in future climate scenarios "Higher Intensity Rainfall" or "Higher Total Rainfall" for the Combined Source Reduction Strategy (not shown). By contrast, cost-effectiveness is often higher than $0.35 per gallon when assuming 2003 Typical Year rainfall (Figure 5.7, top pane), except in cases in which average GSI cost per impervious acre is less than $324,000. In the Recent Historical scenario (Figure 5.7, bottom pane), the results are mixed: If GSI capital costs are higher than $415,000 per acre or the percentage of pipe needing repair is greater than 76 percent, cost-effectiveness vulnerability typically emerges; but, in other scenarios (yellow shading), cost-effectiveness results generally remain lower than $0.35 per gallon.

---

[8]   Using PRIM, the team identified two relevant regions for scenario discovery for Strategy 27. The first includes 2003 Typical Year hydrology and the cost threshold identified in Figure 5.7, while the second includes Recent Historical (2004–2013) hydrology and the GSI cost and pipe repair thresholds also shown on the figure. For the investigation focusing on Recent Historical hydrology, note that we used PRIM to search for acceptable cases rather than vulnerable ones to improve algorithm performance. As a result, any scenario above *either* threshold, rather than both thresholds together, would likely lead to poor cost-effectiveness performance. The unshaded portions of Figure 5.7 show the resulting regions from this multistep investigation. These two regions together capture 77 percent of the scenarios with cost-effectiveness higher than $0.35 per gallon (coverage), and 69 percent of scenarios with these characteristics yield above-threshold cost-effectiveness (density).

Figure 5.7
Scenario Discovery Results for a Combined Source Reduction Strategy (Strategy 27)

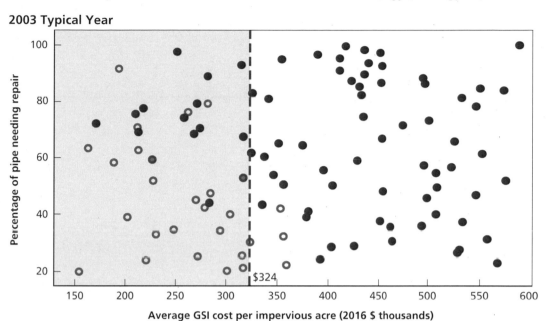

NOTE: Each point summarized represents one mapping of assumptions to consequence, and the points are not assumed to be equally likely. Some points may overlap. Unshaded regions are identified as vulnerable in the scenario discovery analysis, while shaded regions are not vulnerable. Dashed red lines indicate thresholds identified using the PRIM algorithm.

As with GSI-20, these results show that the Combined Source Reduction Strategy is increasingly cost-effective as average annual rainfall increases and the climate scenario becomes more adverse. It also results in acceptable cost-effectiveness assuming either 2003 Typical Year or Recent Historical rainfall if GSI capital costs can be kept in the bottom third to half of the current plausible range and pipe repair programs can be optimized and targeted to the highest inflow pipes while repairing less than three-quarters of the total.

The results also help to clarify the uncertain drivers that do not substantially affect cost-effectiveness performance. Land use and a range of other cost uncertainty dimensions also varied in the analysis, including pipe repair costs, percentage of manholes repaired, and treatment or interceptor cleaning costs (Table 5.3), and do not meaningfully influence the identified scenario regions. This can help Pittsburgh and other municipalities in the ALCOSAN service area focus on key drivers to better address uncertainty in subsequent sewer overflow source reduction planning and design efforts. It also helps to establish thresholds for rainfall monitoring: If monitored rainfall volumes are increasing and begin to resemble the future climate scenarios considered here, this could signal that additional source reduction investments would reduce overflow more cost-effectively.

In general, this pilot study provides initial evidence that broader investments in GSI and I&I source reduction, coupled with gray infrastructure improvements, could yield cost-effective overflow reduction (depending on rainfall, cost, and implementation assumptions). In addition, it suggests that source reduction could provide hedging value for ALCOSAN and municipal planners against future climate and hydrology change. If stormwater source reduction approaches can yield additional sewer overflow reduction in scenarios with more rainfall, they could yield long-term benefits and investment value. Conversely, if rainfall volumes are not increasing or cost and performance targets cannot be achieved, source reduction might not "pay off" well when compared with other infrastructure investments.

Conclusion

This chapter describes an initial investigation of stormwater source reduction and overflow reduction strategies for communities in the ALCOSAN service area across a range of future uncertainties. RDM is designed as an iterative process, and the analysis described in this chapter represents the first pass through the key analysis steps. Critically, we found that none of the strategies initially evaluated achieves the goal of complete or near-complete elimination of SSOs and substantial reduction of CSOs, even in the least-stressing scenarios considered. As a result, none of the strategies considered here represents a potential full solution to this significant challenge, and more investigation is needed to understand how the policy levers evaluated could be coupled with additional infrastructure investments to reliably eliminate sewer overflows.

However, this analysis does provide evidence to help highlight some near-term, low-regret investments—the first step toward an adaptive strategy. First, it confirms that upgrading the WWTP—an early action step requested by USEPA during recent negotiations (see Chapter Two)—can yield significant and cost-effective overflow reduction, and this appears to be a low-regret choice for the near term. Based on a preliminary cost estimate range, the simulation results also show that capital investments to clean the main interceptors could also represent a worthwhile near-term, low-regret option.

By contrast, investing in GSI source reduction to control runoff from up to 20 percent of impervious cover in the combined sewer system, when evaluated in isolation, yielded poor cost-effectiveness for overflow reduction under current rainfall, GSI performance, and capital cost assumptions. These results suggest that GSI investments may not be justifiable for sewer overflow reduction alone if uncertainty is not considered. However, when uncertainty is included, scenarios in which GSI is cost-effective begin to emerge, especially with higher future rainfall. In addition, considering additional GSI co-benefits not addressed here, such as flood risk reduction, nutrient removal, property value increases, or ecosystem services, could still make these investments justifiable in terms of overall societal benefit, even if not justified for overflow reduction alone.

Next, with the treatment expansion and interceptor cleaning implemented, investments in source reduction—including controlling up to 20 percent of impervious area, a systematic pipe repair program focused on high-inflow areas of the separated sewer system, or combinations of both—appear to yield notable overflow reduction benefits that can be cost-effective under plausible future assumptions. Cost-effectiveness of treatment expansion combined with source reduction appear to increase rather than decline in scenarios with higher average rainfall, so source reduction could also help ALCOSAN and municipal planners hedge against more adverse climate scenarios that may exceed the design criteria of existing or proposed gray infrastructure. If cost-effectiveness is a concern, larger investments in source reduction could be included as an adaptive option, triggered in the future after careful monitoring of average annual rainfall patterns to assess the likelihood of higher-rainfall scenarios.

Of course, caution is needed in interpreting these results. In this pilot study, we evaluated an initial set of planning-level strategies that did not include all proposed levers, including key components of the draft WWP, such as deep interceptor storage tunnels or new conveyance pipes. In addition, we compared results across a range of plausible scenarios, but some uncertainties identified by study participants (see Chapter Two) could not be included because of the level of effort or computing cost constraints. Similarly, source reduction policy levers were simulated using simplified, planning-level assumptions. A more-detailed simulation of GSI or I&I source reduction reflecting potential implementation constraints, or further optimized design, could yield either lower or higher performance than what was initially simulated. Finally, this analysis does not take into account the timing or phasing of proposed investments, expected service life, detailed life-cycle costs (such as O&M), or ratepayer affordability, all of which limits the utility of the strategy comparisons and cost-effectiveness estimates. Additional iterations of analysis could help address these limitations and help ALCOSAN, the City of Pittsburgh, and other municipalities identify a more complete long-term adaptive strategy for overflow reduction and stormwater management.

CHAPTER SIX

Key Findings and Next Steps

In the final chapter of this report, we provide a summary of the key findings from this pilot study and next steps in analysis to support planning and management. The chapter ends with a brief discussion of the limitations of this pilot study.

Key Findings

Simulations of the recent past suggest that overflow volumes are up to 15 percent higher than previously estimated. An early step in this investigation was applying a series of simulation models to estimate CSOs and SSOs under Recent Historical rainfall conditions (2004 to 2013). Results suggest that the overflow challenge may have grown in the past decade, with sewer overflows increasing from approximately 9.5 Bgal./year in a 2003 Typical Year simulation to a ten-year average of 11 Bgal./year for 2004 to 2013, when holding other system characteristics constant. These increases could be due, in part, to an increase in average annual rainfall but likely also reflects differences in storm patterns and intensity when comparing the recent ten-year period with a single average or "typical" year initially developed to represent 1948–2008 hydrology. Notably, we see a potential upward trend in average annual rainfall for the region in the past 60 or more years, as well as significant year-to-year variations in the amount of rainfall observed from 2004 through 2014, resulting in annual overflow volumes ranging from roughly 9 Bgal./year to more than 15 Bgal./year.

Future rainfall, population, and land-use changes could increase overflow volumes. Using the same set of simulation models, the RAND team conducted a series of quantitative experiments using high-performance computing to explore how the system as currently constructed and operated might respond to future change. We considered plausible future effects on sewer overflows from several key uncertain factors identified by participating partners and stakeholders, including climate change, future wastewater customer connections, and land-use changes in the combined sewer area. A total of 18 scenarios were simulated, combining different assumptions for each of these factors.

Each of the uncertain factors contributed to overflow increases in the simulation analysis, but no single driver emerged as dominant. The extent of increased vulnerability depends on the assumptions made but ranges from 1.5 to 4.2 Bgal./year in additional overflow volume (11 to 13.7 Bgal./year in total) when compared with a 2003 Typical Year simulation (9.5 Bgal./year). Of this total, SSOs increased to between 0.8 and 1.1 Bgal./year in the future simulations with no additional policy action.

This analysis included a high-population scenario (2xPGH), in which the population in Pittsburgh would nearly double by 2046. This leads to a 24-percent increase in impervious cover for the area contributing directly to CSOs (combined sewer area; see Figure S.1 in the Summary). The corresponding increase in system overflows—roughly 10 percent—was relatively modest. The results suggest that the system is less vulnerable to future impervious growth in the combined area. In other words, although the system faces substantial challenges under present conditions, the combined area appears to already be substantially "built out" at present and there is less risk associated with future population growth. This also might lead local planners to prioritize land-use planning for existing impervious cover in the combined sewer area over addressing potential new growth.

Expanding WWTP capacity or cleaning existing deep interceptors could represent low-regret, near-term options. Next, we identified and tested a range of proposed planning-level strategies intended to either improve the function of the existing sewer system (e.g., expanding the capacity of the regional WWTP) or reduce the flow of stormwater into the system during rainfall events. We conducted a preliminary screening analysis of 30 planning-level combinations of near-term gray infrastructure improvements, I&I reduction (pipe repair), and GSI using a single set of assumptions and then used RDM techniques to evaluate a subset of nine promising strategies identified during screening across a range of nearly 5,000 uncertain scenarios. Uncertain factors considered included climate, wastewater customer connections, and land use, along with additional uncertainties related to strategy cost and GSI performance.

This analysis shows that upgrading the WWTP can yield substantial and cost-effective overflow reduction, and this policy lever appears to be a low-regret choice for near-term implementation. Depending on future assumptions, expanding ALCO-SAN's single treatment plant to 480 MGD in isolation reduces overflows 2.5 to 3.7 Bgal./year in the simulated results, and the corresponding cost-effectiveness ranges from $0.06 per gallon to $0.20 per gallon of overflow reduced. Similarly, the simulation results also show that capital investments to clean the existing deep-tunnel interceptors and expand the treatment plant could also represent a worthwhile near-term, low-regret option across a range of uncertain cost assumptions.

Strategies combining treatment expansion, interceptor cleaning, and source reduction could cost-effectively reduce future overflow, but uncertainty remains high. Simulation analysis results also suggest that combining a WWTP expansion with incremental investments in I&I or GSI source reduction could yield substantial and cost-effective overflow reduction under plausible future assumptions. However, the uncertainty associated with both overflow reduction performance and source reduction costs remains high, with corresponding uncertainty about cost-effectiveness. For instance, the Combined Source Reduction Strategy (Strategy 27), which includes a treatment plant expansion, existing deep-tunnel interceptor cleaning, GSI investments intended to control runoff from 20 percent of impervious cover in the combined sewer area (GSI-20), and I&I reduction of 20 to 40 percent for targeted areas of high inflow (I&I Mid [8%]), yields overflow reduction of 4.6 to 7.1 Bgal./year across the range of scenarios considered. Estimated capital costs for this strategy vary from $816 million to $3.1 billion, yielding a range of cost-effectiveness from $0.11 per gallon to $0.67 per gallon.

None of the strategies combining treatment plant expansion with interceptor cleaning and source reduction fully eliminates sewer overflows. This analysis tested a wide range of proposed stormwater source reduction strategies—that is, strategies that relying on reducing stormwater inflows into the sewer system through GSI or pipe repair—including those with investments in GSI that could be beyond currently plausible implementation limits (e.g., GSI-40 intended to control runoff from up to 40 percent of all impervious areas). Even when simulating these "art of the possible" strategies, however, we found that no single or combined strategy eliminates SSOs (a requirement of the PADEP CD) or nearly eliminates CSOs in any scenario, even in the least stressing scenarios considered. For instance, Strategy 29, which includes all policy levers considered and GSI-40, reduces SSOs 37 to 53 percent but still produces 0.3 to 0.5 Bgal./year of SSOs (and 3.8 to 6.4 Bgal./year of total remaining overflow) across the simulated scenarios.

As a result, none of the planning-level strategies considered in this report represents a potential full solution to this significant challenge, and more investigation is needed to understand how the evaluated policy levers could be combined with additional infrastructure investments—such as the new conveyance, storage tunnels, or other improvements identified in ALCOSAN's draft WWP—to reliably eliminate sewer overflows as part of a long-term strategy.

Source reduction strategies are more cost-effective in higher-rainfall scenarios. As a final step in the RDM analysis, we used scenario discovery methods to help identify the key uncertain drivers for cost-effectiveness of either one GSI-only strategy (GSI-20; Strategy 2) or for one possible Combined Source Reduction Strategy (Strategy 27). Using this process, we sought to identify uncertain drivers that yield cost-effectiveness either lower than (acceptable) or higher than (poor) a threshold of $0.35 per gallon.

For GSI-20 alone (Strategy 2), key uncertain drivers identified include future climate, GSI performance, and average GSI cost per impervious acre controlled. Under historical climate assumptions, GSI-20 yields acceptable cost-effectiveness only with High GSI performance and relatively low average per-acre capital costs (less than

$230,000 per acre controlled). In other words, under existing rainfall conditions, one has to carefully manage GSI costs and identify high-flow sites to achieve cost-effective performance. In plausible future climate scenarios with higher rainfall volumes, however, GSI-20 can be cost-effective assuming "Low" GSI performance and across a wider range of capital costs overall ($283,000 to $353,000).

Rainfall uncertainty is also a key driver for a Combined Source Reduction Strategy. As with Strategy 2, this strategy is increasingly cost-effective as average annual rainfall increases and the climate scenario becomes more adverse. Strategy 27 simulation results show acceptable cost-effectiveness in nearly all scenarios that include plausible future increases in annual rainfall. Under historical climate, the strategy is cost-effective when the average GSI cost per impervious acre is less than $324,000 (2003 Typical Year rainfall) or the GSI cost per acre is less than $415,000 and less than 76 percent of pipes are assumed to need repair to achieve the I&I reduction target (Recent Historical rainfall). Put another way, poor cost-effectiveness performance for the Combined Source Reduction Strategy emerges only through a combination of assumptions, including non-increasing rainfall and either higher GSI per-acre costs or a high percentage of pipes needing repair.

This analysis shows that investing in GSI source reduction to control up to 20 percent of impervious cover in the combined sewer system, when evaluated in isolation, yields poor cost-effectiveness for overflow reduction under "nominal" or commonly used rainfall, GSI performance, and capital cost assumptions. The results suggest that GSI investments may not be justifiable for sewer overflow reduction alone if uncertainty is not considered.

However, taking into account plausible future change, the analysis also shows that source reduction investments similar to those evaluated here could provide long-term value, even if their investment performance compares unfavorably to gray infrastructure under current conditions or historical hydrology alone. Specifically, source reduction could provide near-term hedging value for ALCOSAN and municipal planners against future climate and hydrology changes. If increases in average annual rainfall appear more likely, this could make these source reduction approaches more cost-effective to implement in the long term. By contrast, the performance of gray infrastructure designed for a specific hydrology or rainfall assumption might decline if these design assumptions are exceeded, leading to poorer performance and lower cost-effectiveness. Source reduction investments also provide flexibility in terms of the timing and sequence of project implementation, as well as the potential for incremental benefits over time, when compared with large-scale gray infrastructure.

Next Steps for Stormwater Analysis to Support Planning

Plausible future change should inform near-term planning and design. The vulnerability analysis described in Chapter Three suggests that the sewer overflow challenge may have already grown in recent years and could increase further under plausible assumptions about future climate, population growth, or land-use changes. Infrastructure planning and design based only on a 2003 Typical Year could yield a system that is not resilient to these changes, meaning that overflows could still occur regularly even after these investments are implemented. Related cost estimates may also be too low—if system components need to be sized to account for additional rainfall, for example, long-term implementation costs may end up higher than currently projected. These considerations suggest a need to incorporate a range of assumptions about future rainfall, project costs, and GSI performance when planning for or designing key components of the draft WWP, identifying source reduction investments for municipalities across the ALCOSAN service area, and designing and implementing a complete green infrastructure strategy for Pittsburgh.

Source reduction should be considered for benefits beyond sewer overflow reduction. This pilot study began with a series of scoping workshops, in which participants identified a wide-ranging and ambitious research agenda. Participants collectively identified what could be considered the scope of an integrated regional watershed management planning effort, supported by uncertainty analysis and scenario planning, for Allegheny County watersheds. This report addresses sewer overflows and water-quality planning, one of the central concerns for regional stormwater management identified by study participants. The analysis shows that large-scale investments in source reduction could help reduce overflow but with a wide range of uncertainties regarding cost-effectiveness and relative strategy performance. The analysis showed that source reduction was generally a less cost-effective approach to reducing overflows than the other policy levers considered.

However, participants in the workshops also identified such goals as flood risk reduction, ecosystem services, access to green space, and economic development. Stormwater source reduction and, in particular, GSI are often encouraged to help address these goals alongside water-quality improvement, but we were unable to evaluate these important co-benefits within the scope of this study.

Future research could build on the RDM uncertainty analysis approach described here while also incorporating estimates of source reduction co-benefits—which may include health, environmental, and economic effects—to provide a more-complete understanding of the benefits and costs of source reduction as part of regional, integrated stormwater management plans. In turn, this could help ALCOSAN, the City of Pittsburgh, and other municipalities identify solutions that are more cost-effective and yield a broader range of benefits. The analysis framework described in this report could, in general, help support continued progress toward regional collaboration on stormwater planning.

Source reduction could help reliably reduce overflows, but additional research is needed to fully define a long-term, adaptive strategy. This analysis suggests that investments in GSI and I&I source reduction, coupled with treatment plant expansion, could yield cost-effective overflow reduction if future rainfall volumes increase and certain cost and performance assumptions can be realized. In addition, it suggests that investment in regional source reduction could provide hedging value for ALCOSAN and municipal planners against future rainfall increases, potentially avoiding the need to further upgrade the gray infrastructure system. These preliminary findings should be followed by further analysis.

As a pilot effort, our study represents a first iteration of RDM analysis for regional stormwater management, intended to estimate future vulnerability for the current system and begin to evaluate the benefits and costs from regional source reduction strategies. None of the strategies considered was able to meet overflow reduction goals in any of the simulated scenarios; however, an important next step would be to apply a similar robustness and deep uncertainty framework with an expanded range of policy levers and strategies. For instance, additional components of the draft WWP—such as further expansion of the treatment plant, new deep-tunnel interceptors, or other new conveyance and storage not evaluated in this research scope—could be combined with I&I or GSI source reduction levers to determine the level of investment needed to reliably reduce or eliminate overflows both today and decades from now.

Additional analysis using deep uncertainty planning methods could identify specific phasing for gray and green investments as part of an adaptive plan. Such a plan might include additional low-regret, near-term options; defer some high-cost investment decisions—for example, the proposed new storage tunnels along the major rivers—to a later date; and incorporate monitoring for key uncertain factors (e.g., average annual rainfall, GSI per-acre costs) to determine when these additional investments should be made. Research focused on resolving existing data gaps surrounding project costs—for instance, bounding and quantifying long-term O&M costs for GSI of different types—would also be an important step to better compare and identify robust options moving forward. Our analysis is suggestive, but not yet conclusive, about what a full adaptive strategy might entail.

Limitations of This Study

This pilot study represents an important step forward for incorporating and addressing uncertainty in stormwater and combined sewer management for Allegheny County. Some notes of caution are warranted.

This pilot may not capture the full range of uncertainty associated with current system vulnerability or future strategy performance. Our results are based on a limited number of uncertain factors and scenarios evaluated. Although we modeled more than 500 unique scenario and year combinations throughout this analysis, which required more than 30,000 CPU-hours of computing time, we were limited by the cost of computing resources to explore additional scenarios and strategies in some cases. RDM and related deep uncertainty methods often call for systematic testing across much larger numbers of plausible scenarios, however, and including additional uncertain factors or scenarios related to sewer overflow could simulate a wider range of uncertainty.

Of particular note were the simplifying assumptions needed to evaluate stormwater source reduction at this early stage. We set out to quantify the potential benefits and costs of source reduction while accounting for future climate and other uncertainty but, in doing so, relied on assumptions regarding strategy implementation. Specifically, this report does not address the barriers to implementation currently faced by ALCOSAN and municipal planners for large-scale GSI or I&I source reduction and assumes that source reduction could be implemented as represented in the simulation models. These potential barriers, which include capital and financing availability and fragmented infrastructure ownership and regional decisionmaking, are crucial to the success or failure of future stormwater investments in the region, but they are not yet incorporated into this decision framework.

In addition, source reduction policy levers were also represented in the SWMM modeling using simplified, planning-level assumptions. A more-detailed investigation of GSI or I&I source reduction, taking into account site availability, feasibility, land ownership, or other constraints, could yield either lower or higher performance than what was initially simulated. These additional implementation and performance uncertainties could be included in future iterations of uncertainty analysis, however, building on the same framework and approach described in this report. As noted in the previous section, potential co-benefits from GSI were also not considered in this pilot analysis. These co-benefits, if included, could significantly influence the comparative ranking and overall cost-effectiveness of proposed stormwater management strategies.

Finally, as noted earlier, we evaluated an initial set of planning-level strategies that did not encompass all proposed levers, including such key components of the draft WWP as deep interceptor storage tunnels or new conveyance pipes. The capital costs to clean existing interceptors are based on preliminary estimates only and have not yet been formally scoped. In addition, strategy cost comparisons do not yet take into account the timing or phasing of proposed investments, the expected service life of projects, O&M or detailed project life-cycle costs, or ratepayer affordability—all of which limit the utility of the strategy comparisons and cost-effectiveness estimates. These could be significant factors, because, for example, uncertainty related to the cost and effort needed for GSI maintenance remains high. The timing or phasing of different proposed policy levers was also not considered in the strategy comparison.

Many of these limitations could be addressed through future iterations of analysis building on this decision analysis framework, however. Such additional steps could help municipalities in Allegheny County move toward more adaptive and resilient long-term stormwater planning and management.

Abbreviations

2xPGH	high growth (one of three land-use scenarios)
ACE	Allegheny County Executive
ACHD	Allegheny County Health Department
ACT	Alternatives Costing Tool
ALCOSAN	Allegheny County Sanitary Authority
AOP	Art of the Possible
AWS	Amazon Web Services
Bgal.	billion gallons
CC	Chartiers Creek
CD	consent decree
CONNECT	Congress of Neighboring Community
CPU	central processing unit
CSO	combined sewer overflow
CWA	Clean Water Act
DCIA	directly connected impervious area
FWOA	future without action
GCM	General Circulation Model
GFDL	Geophysical Fluid Dynamics Laboratory General Circulation Model
GIS	Geographic Information Systems
GR	Girty's Run
GSI	green stormwater infrastructure
GWI	groundwater inflows
H&H	hydrologic and hydraulic
HadCM3	Hadley Center Coupled General Circulation Model, version 3

HGL	hydraulic grade line
HI	High Infiltration
HRM3	Hadley Regional Model 3 Regional Climate Model
I&I	inflow and infiltration
IQR	interquartile range
IRO	interceptor relief overflow
LHS	Latin hypercube sampling
LO	Lower Ohio
LTCP	long-term control plan
Mgal.	million gallons
MGD	million gallons per day
MM5I	Penn State University/National Center for Atmospheric Research Mesoscale Regional Climate Model
MR	Main Rivers
NARCCAP	North American Regional Climate Change Assessment Program
O&M	operations and maintenance
PADEP	Pennsylvania Department of Environmental Protection
PRIM	Patient Rule Induction Method
PWSA	Pittsburgh Water and Sewer Authority
QQM	quantile-quantile mapping
RBM	Regional Balance Model
RCM	Regional Climate Model
RDII	rainfall-derived inflow and infiltration
RDM	Robust Decision Making
SM	Saw Mill Run
SMP	Stormwater Management Plan
SPC	Southwestern Pennsylvania Commission
SRES	Special Report on Emissions Scenarios
SRIC	Sewer Regionalization Implementation Committee
SSO	sanitary sewer overflow
SWMM	Storm Water Management Model
TC	Turtle Creek
UA	Upper Allegheny

UM	Upper Monongahela
USEPA	U.S. Environmental Protection Agency
WPAC	Watershed Plan Advisory Committee
WWP	Wet Weather Plan

References

3 Rivers Wet Weather, homepage, undated-a. As of March 20, 2017:
http://www.3riverswetweather.org/

———, "About the Wet Weather Issue," web page, undated-b. As of March 20, 2017:
http://www.3riverswetweather.org/about-wet-weather-issue

———, "Green Solutions," web page, undated c. As of March 20, 2017:
http://www.3riverswetweather.org/green/green-solutions

———, "Green Solutions: Rain Garden," web page, undated-d. As of March 20, 2017:
http://www.3riverswetweather.org/green/green-solution-rain-garden

———, "History," web page, undated-e. As of March 20, 2017:
http://www.3riverswetweather.org/about-us/how-program-operates/history-mission

———, "Understanding the Sewer Collection System: History," web page, undated-f. As of March 20, 2017:
http://www.3riverswetweather.org/about-wet-weather-issue/understanding-sewer-collection-system/history

ACHD—*See* Allegheny County Health Department.

ALCOSAN—*See* Allegheny County Sanitary Authority.

Allegheny County Health Department, "Article XIV–Sewage Disposal Rules and Regulations," December 1, 1997. As of March 21, 2017:
http://www.achd.net/waterw/pubs/pdf/sewage.pdf

Allegheny County Sanitary Authority, "Green First Update: Reducing Overflows at the Source," web page, undated. As of April 13, 2017:
http://www.alcosan.org/GreenFirstUpdate/tabid/208/Default.aspx

———, "1.4 Wet Weather Plan Development Process and Planning Team," *ALCOSAN's Draft Wet Weather Plan*, Pittsburgh, Pa., 2012a. As of March 17, 2017:
http://www.alcosan.org/Portals/0/Wet%20Weather%20Plan/Section%201.pdf

———, "3.0 Existing Systems and Conditions," *ALCOSAN's Draft Wet Weather Plan*, Pittsburgh, Pa., 2012b. As of March 17, 2017:
http://www.alcosan.org/Portals/0/Wet%20Weather%20Plan/Section%203.1.pdf

———, "4.0 Hydrologic and Hydraulic Characterization," *ALCOSAN's Draft Wet Weather Plan*, Pittsburgh, Pa., 2012c. As of March 17, 2017:
http://www.alcosan.org/Portals/0/Wet%20Weather%20Plan/Section%204.0%20thru%204.4.pdf

———, "9.0 Alternatives Analysis," *ALCOSAN's Draft Wet Weather Plan*, Pittsburgh, Pa., 2012d. As of March 17, 2017:
http://www.alcosan.org/Portals/0/Wet%20Weather%20Plan/Section%209.1.pdf

————, "10.0 Recommended 2026 Plan (Part 1)," *ALCOSAN's Draft Wet Weather Plan*, Pittsburgh, Pa., 2012e. As of March 17, 2017:
http://www.alcosan.org/Portals/0/Wet%20Weather%20Plan/Section%2010.0%20thru10.6.pdf

————, "11.0 Implementation Plan," *ALCOSAN's Draft Wet Weather Plan*, Pittsburgh, Pa., 2012f. As of March 17, 2017:
http://www.alcosan.org/Portals/0/Wet%20Weather%20Plan/Section%2011.pdf

————, *ALCOSAN's Draft Wet Weather Plan*, Pittsburgh, Pa., 2012g. As of March 17, 2017:
http://www.alcosan.org/WetWeatherIssues/ALCOSANDraftWetWeatherPlan/
DraftWWPFullDocument/tabid/176/Default.aspx

————, "Appendix C: GSI Cost Literature Review," *Starting at the Source: How Our Region Can Work Together for Clean Water*, Pittsburgh, Pa., 2015a.

————, *Starting at the Source: How Our Region Can Work Together for Clean Water*, Pittsburgh, Pa., 2015b.

Allegheny Conference on Community Development, *Sewer Regionalization Evaluation: Review Panel Findings and Recommendations*, Pittsburgh, Pa., March 15, 2013. As of March 17, 2017:
http://www.alcosan.org/Portals/0/PDFs/
SEWER%20REGIONALIZATION%20EVALUATION%20REPORT_MARCH%202013[1].pdf

Allegheny County, *Home Rule Charter of Allegheny County*, Chapter C, January 2000. As of March 20, 2017:
http://ecode360.com/8453332

Allegheny County Health Department, "Article XIV–Sewage Disposal Rules and Regulations," December 1, 1997. As of March 21, 2017:
http://www.achd.net/waterw/pubs/pdf/sewage.pdf

Allegheny Places, "Supporting Documentation: Act 167 Stormwater Management Planning," *The Allegheny County Comprehensive Plan*, undated. As of March 21, 2017:
http://www.alleghenyplaces.com/Comprehensive_plan/Sup_Doc.aspx

Arnbjerg-Nielsen, K., P. Willems, K. Olsson, S. Beecham, A. Pathirana, I. Bulow-Gregersen, H. Madsen, and V. Ngyuen, "Impacts of Climate Change on Rainfall Extremes and Urban Drainage Systems: A Review," *Water Science and Technology: A Journal of the International Association on Water Pollution Research*, Vol. 68, No. 1, 2013, pp. 16–28.

Balingit, M., "Fatal Flooding in 2011 on Washington Boulevard Brings Suit Against Many," *Pittsburgh Post-Gazette*, February 2, 2013. As of March 21, 2017:
http://www.post-gazette.com/local/city/2013/02/02/
Fatal-flooding-in-2011-on-Washington-Boulevard-brings-suit-against-many/stories/201302020263

Bankes, S. C., "Exploratory Modeling for Policy Analysis," *Operations Research*, Vol. 41, No. 3, 1993, pp. 435–449.

Boe, J., L. Terray, F. Habets, and E. Martin, "Statistical and Dynamical Downscaling of the Seine Basin Climate for Hydro-Meteorological Studies," *International Journal of Climatology*, Vol. 27, No. 12, 2007, pp. 1643–1656.

Brown, C., Y. Ghile, M. Laverty, and K. Li, "Decision Scaling: Linking Bottom-Up Vulnerability Analysis with Climate Projections in the Water Sector," *Water Resources Research*, Vol. 48, No. 9, 2012.

Bryant, B. P., and R. J. Lempert, "Thinking Inside the Box: A Participatory Computer-Assisted Approach to Scenario Discovery," *Technological Forecasting and Social Change*, Vol. 77, No. 1, 2010, pp. 34–49.

Burian, S., S. Nix, R. Pitt, and S. Durrans, "Urban Wastewater Management in the United States: Past, Present, and Future," *Journal of Urban Technology*, Vol. 7, No. 3, 2000, pp. 33–62.

Center for Watershed Protection, photo of permeable pavement, New York State Stormwater Flickr, 2015.

City of Milwaukee, "Stormwater and Sewer Capacity," undated. As of March 21, 2017:
http://city.milwaukee.gov/commoncouncil/District10/
Stormwater-and-Sewer-Capacity.htm#.V9mN6xRu7iQ

Collins, T., J. Kline, K. Vallianos, and C. Fox, *Ecology and Recovery: Allegheny County*, Pittsburgh, Pa.: Studio for Creative Inquiry, Carnegie Mellon University, 2005. As of March 20, 2017:
http://3r2n.collinsandgoto.com/revalued/ecology-recovery-allegheny-county/index.htm

Comebemale, C., J. Jels-Seale, S. Laventure-Volz, and Q. Yu, *Pathways to Collaboration: Municipal Decision Making and Stormwater Management in Allegheny County*, Pittsburgh, Pa.: Heinz College System Synthesis Report, Carnegie Mellon University, 2016.

Committee on Adaptation to a Changing Climate, *Adapting Infrastructure and Civil Engineering Practice to a Changing Climate*, J. R. Olsen, ed., Reston, Va.: American Society of Civil Engineers, 2015.

Commonwealth of Pennsylvania, Department of Environmental Protection, Storm Water Management Act, Public Law 864, Act 167, October 4, 1978, last revised 2008.

Congressional Budget Office, *Public Spending on Transportation and Water Infrastructure, 1956 to 2014*, Washington, D.C., March 2015. As of March 20, 2017:
https://www.cbo.gov/publication/49910

Cook, Louis, photo of vegetative swale, Philadelphia Water Department, 2013.

David L. Lawrence Convention Center, photo of green roof, Flickr, 2013.

District of Columbia Water and Sewer Authority, "Long Term Control Plan Modification for Green Infrastructure," May 2015. As of March 20, 2017:
https://www.dcwater.com/sites/default/files/green-infrastructure-ltcp-modificaitons.pdf

Exum, L., S. Bird, J. Harrison, and C. Perkins, *Estimating and Projecting Impervious Cover in the Southeastern United States*, Athens, Ga.: U.S. Environmental Protection Agency, Ecosystems Research Division, EPA/600/R-05/061, 2005.

Fischbach, Jordan R., *Managing New Orleans Flood Risk in an Uncertain Future Using Non-Structural Risk Mitigation*, Santa Monica, Calif.: RAND Corporation, RGSD-262, 2010. As of March 15, 2017:
http://www.rand.org/pubs/rgs_dissertations/RGSD262.html

Fischbach, Jordan R., Robert J. Lempert, Edmundo Molina-Perez, Abdul Ahad Tariq, Melissa L. Finucane, and Frauke Hoss, *Managing Water Quality in the Face of Uncertainty: A Robust Decision Making Demonstration for EPA's National Water Program*, Santa Monica, Calif.: RAND Corporation, RR-720-EPA, 2015. As of March 15, 2017:
http://www.rand.org/pubs/research_reports/RR720.html

Gaffin, S. R., C. Rosenzweig, and A. Kong, "Adapting to Climate Change Through Urban Green Infrastructure," *Nature Climate Change*, Vol. 2, No. 10, 2012, p. 704.

Gill, S., J. Handley, A. Ennos, and S. Pauleit, "Adapting Cities for Climate Change: The Role of the Green Infrastructure," *Built Environment*, Vol. 33, No. 1, 2007, pp. 115–133.

Groves, D. G., E. Bloom, R. J. Lempert, J. R. Fischbach, J. Nevills, and B. Goshi, "Developing Key Indicators for Adaptive Water Planning," *Journal of Water Resources Planning and Management*, Vol. 141, No. 7, 2015.

Groves, David G., Jordan R. Fischbach, Evan Bloom, Debra Knopman, and Ryan Keefe, *Adapting to a Changing Colorado River: Making Future Water Deliveries More Reliable Through Robust Management Strategies*, Santa Monica, Calif.: RAND Corporation, RR-242-BOR, 2013. As of March 15, 2017:
http://www.rand.org/pubs/research_reports/RR242.html

Groves, David G., Jordan R. Fischbach, Nidhi Kalra, Edmundo Molina-Perez, David Yates, David Purkey, Amanda Fencl, Vishal K. Mehta, Ben Wright, and Grantley Pyke, *Developing Robust Strategies for Climate Change and Other Risks: A Water Utility Framework*, Santa Monica, Calif.: RAND Corporation, RR-977-WRF, 2014. As of March 15, 2017:
http://www.rand.org/pubs/research_reports/RR977.html

Groves, David G., Jordan R. Fischbach, Debra Knopman, David R. Johnson, and Katheryn Giglio, *Strengthening Coastal Planning: How Coastal Regions Could Benefit from Louisiana's Planning and Analysis Framework*, Santa Monica, Calif.: RAND Corporation, RR-437-RC, 2014. As of March 15, 2017:
http://www.rand.org/pubs/research_reports/RR437.html

Groves, D. G., and R. J. Lempert, "A New Analytic Method for Finding Policy-Relevant Scenarios," *Global Environmental Change,* Part A: *Human and Policy Dimensions*, Vol. 17, No. 1, 2007, pp. 73–85.

Gudmundsson, L., J. Bremnes, J. Haugen, and T. Engen-Skaugen, "Technical Note: Downscaling RCM Precipitation to the Station Scale Using Statistical Transformations: A Comparison of Methods," *Hydrology and Earth System Sciences*, Vol. 16, No. 9, 2012, pp. 3383–3390.

Haasnoot, M., J. Kwakkel, W. Walker, and J. ter Maat, "Dynamic Adaptive Policy Pathways: A Method for Crafting Robust Decisions for a Deeply Uncertain World," *Global Environmental Change*, Part A: *Human and Policy Dimensions*, Vol. 23, No. 2, 2013, pp. 485–498.

Hall, J. W., R. J. Lempert, K. Keller, A. Hackbarth, C. Mijere, and D. J. McInerney, "Robust Climate Policies Under Uncertainty: A Comparison of Robust Decision Making and Info-Gap Methods," *Risk Analysis*, Vol. 32, No. 10, 2012, pp. 1657–1672.

Herman, J., P. Reed, H. Zeff, and G. Characklis, "How Should Robustness Be Defined for Water Systems Planning Under Change?" *Journal of Water Resources Planning and Management*, Vol. 141, No. 10, October 2015.

Hopey, D., "EPA Calls ALCOSAN's $2 Billion Sewer System Proposal Deficient," *Pittsburgh Post-Gazette*, January 31, 2014. As of March 20, 2017:
http://www.post-gazette.com/local/city/2014/01/31/
Federal-officials-say-Alcosan-s-sewer-upgrade-plan-doesn-t-go-far-enough/stories/201401310158

———, "ALCOSAN Gets More Time to Create Green Solutions to Overflow Problem," *Pittsburgh Post-Gazette*, March 9, 2016.

Institute for Public Policy and Economic Development, "A Primer on Home Rule," Wilkes-Barre, Pa., 2009. As of March 20, 2017:
http://www.institutepa.org/PDF/Research/APrimeronHomeRule0809.pdf

Kansas City, Missouri, Water Service Department, *Overflow Control Program: Overflow Control Plan*, January 2009, last updated April 30, 2012. As of March 20, 2017:
https://www.kcwaterservices.org/wp-content/uploads/2013/04/
Overflow_Control_Plan_Apri302012_FINAL.pdf

Kleidorfer, M., M. Moderl, R. Sitzenfrei, C. Urich, and W. Rauch, "A Case Independent Approach on the Impact of Climate Change Effects on Combined Sewer System Performance," *Water, Science, and Technology*, Vol. 60, No. 6, 2009, pp. 1555–1564.

Knopman, Debra, and Robert J. Lempert, *Urban Responses to Climate Change: Framework for Decisionmaking and Supporting Indicators*, Santa Monica, Calif.: RAND Corporation, RR-1144-MCF, 2016. As of March 15, 2017:
http://www.rand.org/pubs/research_reports/RR1144.html

Kwadijk, J. C. J., M. Haasnoot, J. P. M. Mulder, M. M. C. Hoogvliet, A. B. M. Jeuken, R. A. A. van der Krogt, N. G. C. van Oostrom, H. A. Schelfhout, E. H. van Velzen, H. van Waveren, and M. J. M. de Wit, "Using Adaptation Tipping Points to Prepare for Climate Change and Sea Level Rise: A Case Study in the Netherlands," *Wiley Interdisciplinary Reviews: Climate Change*, Vol. 1, No. 5, 2010, pp. 729–740.

Lempert, R. J., and M. Collins, "Managing the Risk of Uncertain Threshold Responses: Comparison of Robust, Optimum, and Precautionary Approaches," *Risk Analysis*, Vol. 27, No. 4, 2007, pp. 1009–1026.

Lempert, Robert J., Steven W. Popper, and Steven C. Bankes, *Shaping the Next One Hundred Years: New Methods for Quantitative, Long-Term Policy Analysis*, Santa Monica, Calif.: RAND Corporation, MR-1626-RPC, 2003. As of March 15, 2017:
http://www.rand.org/pubs/monograph_reports/MR1626.html

Mamo, T., "Evaluation of the Potential Impact of Rainfall Intensity Variation Due to Climate Change on Existing Drainage Infrastructure," *Journal of Irrigation and Drainage Engineering*, Vol. 141, No. 10, October 2015.

Mark the Trigeek, photo of rain barrel, Creative Commons, 2011.

Mearns, L., *The North American Regional Climate Change Assessment Program Dataset*, Boulder, Colo.: National Center for Atmospheric Research Earth System Grid data portal, 2007, last updated 2014. As of March 20, 2017:
https://www.earthsystemgrid.org/project/NARCCAP.html

Mearns, L., W. Gutowski, R. Jones, L. Y. Leung, S. McGinnis, A. Nunes, and Y. Qian, "A Regional Climate Change Assessment Program for North America," *EOS*, Vol. 90, No. 36, 2009, pp. 311–312.

Melillo, Jerry M., Terese (T. C.) Richmond, and Gary W. Yohe, eds., *Climate Change Impacts in the United States: The Third National Climate Assessment*, Washington, D.C.: U.S. Global Change Research Program, 2014. As of March 20, 2017:
http://nca2014.globalchange.gov/

Michael Baker Jr., Inc., *Allegheny County Stormwater Management Plan Phase 1 Report*, Moon Township, Pa., December 2014.

Milly, P., J. Betancourt, M. Falkenmark, R. Hirsch, Z. W. Kundzewicz, D. Lettenmaier, and R. J. Stouffer, "Stationarity Is Dead: Whither Water Management?" *Science*, Vol. 319, No. 5863, 2008, pp. 573–574.

Milwaukee Metropolitan Sewerage District, *Regional Green Infrastructure Plan*, Milwaukee, Wisc., June 2013. As of March 20, 2017:
http://www.freshcoast740.com/-/media/FreshCoast740/Documents/GI%20Plan/Plan%20docs/MMSDGIP_Final.pdf

Nakicenovic, N., J. Alcamo, G. Davis, B. d. Vries, J. Fenhann, S. Gaffin, K. Gregory, A. Grübler, T. Y. Jung, T. Kram, E. L. L. Rovere, L. Michaelis, S. Mori, T. Morita, W. Pepper, H. Pitcher, L. Price, K. Riahi, A. Roehrl, H. H. Rogner, A. Sankovski, M. Schlesinger, P. Shukla, S. Smith, R. Swart, S. v. Rooijen, N. Victor, Z. Dadi, and Intergovernmental Panel on Climate Change, *Emissions Scenarios*, in N. Nakicenovic and R. Swart, eds., Cambridge, England: Cambridge University Press, 2000.

National Centers for Environmental Information, "U.S. Annual Climatological Summaries," National Oceanic and Atmospheric Administration, NCEI DSI 3220_01, last updated May 5, 2016. As of March 20, 2017:
https://gis.ncdc.noaa.gov/geoportal/catalog/search/resource/details.page?id=gov.noaa.ncdc:C00040

National Research Council, *Regional Cooperation for Water Quality Improvement in Southwestern Pennsylvania*, Washington, D.C.: National Academies Press, 2005. As of March 20, 2017:
http://www.nap.edu/catalog/11196/regional-cooperation-for-water-quality-improvement-in-southwestern-pennsylvania

———, *Informing Decisions in a Changing Climate*, paper presented at the Panel on Strategies and Methods for Climate-Related Decision Support, Washington, D.C.: National Academies Press, 2009. As of March 20, 2017:
https://www.nap.edu/catalog/12626/informing-decisions-in-a-changing-climate

New York City Department of Environmental Protection, "LTCP Frequently Asked Questions," undated. As of March 20, 2017:
http://www.nyc.gov/html/dep/pdf/cso_long_term_control_plan/ltcp_faqs_handout.pdf

———, *NYC Green Infrastructure Plan: A Sustainable Strategy for Clean Waterways*, New York: City of New York, Office of the Mayor, 2010. As of March 20, 2017:
http://www.nyc.gov/html/dep/pdf/green_infrastructure/NYCGreenInfrastructurePlan_LowRes.pdf

Nilsen, V., J. Lier, J. Bjerkholt, and O. Lindolm, "Analyzing Urban Floods and Combined Sewer Overflows in a Changing Climate," *Journal of Water and Climate Change*, Vol. 2, No. 4, 2011, pp. 260–271.

O'Donnell, A., T. Forsell, L. Scott Horsley, N. Kelly, and K. McAllister, "Economics of Green Infrastructure in Adapting to Extreme Precipitation," presentation at the Carolinas Climate Resilience Conference, Charlotte, N.C., April 28–29, 2014. As of March 20, 2017:
http://www.cisa.sc.edu/ccrc/pdfs/2014/Presentations/
ODonnell_Assessing%20Cost%20Effectiveness%20of%20Green%20Infrastructure.pdf

Office of the Mayor William Peduto, "Statement on EPA Correspondence on Wet Weather Plan," press release, Pittsburgh, Pa., March 9, 2016. As of March 20, 2017:
http://pittsburghpa.gov/mayor/release.htm?id=5798

PADEP—*See* Pennsylvania Department of Environmental Protection.

Pennsylvania Department of Environmental Protection, "Interim Municipal Consent Orders," 2015. As of March 20, 2017:
http://www.3riverswetweather.org/municipalities/municipal-consent-orders/
interim-municipal-consent-orders

Philadelphia Water Department, *Philadelphia Combined Sewer Overflow Long Term Control Plan Update, Supplemental Documentation*, Volume 3: *Basis of Cost Opinions*, September 2009. As of March 20, 2017:
http://www.phillywatersheds.org/ltcpu/Vol03_Cost.pdf

———, photo of infiltration trench, Flickr, 2012.

———, "Green City, Clean Waters: Philadelphia's Long-Term Control Plan to Address Combined Sewer Overflows," October 24, 2013. As of March 20, 2017:
http://www.montcopa.org/DocumentCenter/View/6567

Pittsburgh Water and Sewer Authority, "Greening the Pittsburgh Wet Weather Plan," July 2013. As of March 20, 2017:
http://apps.pittsburghpa.gov/pwsa/PWSA-Greening_the_Pittsburgh_Wet_Weather_Plan.pdf

———, "Draft City-Wide Green First Plan," 2016. As of March 20, 2017:
http://pgh2o.com/City-Wide-Green-Plan

PWSA—*See* Pittsburgh Water and Sewer Authority.

River Life Task Force, "A Vision Plan for Pittsburgh's Riverfront," 2001.

Rossman, Lewis, *Stormwater Management Model User's Manual*, Version 5.1, Cincinnati, Ohio: National Risk Management, Laboratory Office of Research and Development, U.S. Environmental Protection Agency, EPA/600-R-14/413b, revised September 2015. As of March 20, 2017: http://nepis.epa.gov/Exe/ZyPDF.cgi?Dockey=P100N3J6.TXT

Savage, L. J., *The Foundations of Statistics*, New York: Wiley Publications, 1954.

Semadeni-Davies, A., C. Hernebring, G. Svensson, and L. Gustafsson, "The Impacts of Climate Change and Urbanisation on Drainage in Helsingborg, Sweden: Combined Sewer System," *Journal of Hydrology*, Vol. 350, No. 1–2, 2008, pp. 100–113.

Sewer Regionalization Implementation Committee, *Sewer Regionalization Implementation Committee Regionalization Update: 2015*, Pittsburgh, Pa., 2015. As of March 20, 2017: http://www.3riverswetweather.org/sites/default/files/
SRIC%20Regionalization%20Update%202015%20final%20report.pdf

Shortle, J., D. Abler, A. Britson, K. Fang, A. Kernanian, P. Knight, M. McDill, R. Najjar, M. Nessry, R. Ready, A. Ross, M. Rydzik, C. Shen, S. Wang, D. Ward, and S. Yetter, *Pennsylvania Climate Impacts Assessment Update*, State College, Pa.: Pennsylvania State University, 2015. As of March 20, 2017:
http://www.elibrary.dep.state.pa.us/dsweb/Get/Document-108470/2700-BK-DEP4494.pdf

Tarr, J. A., and T. F. Yosie, "Critical Decisions in Pittsburgh Water and Wastewater Treatment," in J. A. Tarr, ed., *Devastation and Renewal: An Environmental History of Pittsburgh and Its Region*, Pittsburgh, Pa.: University of Pittsburgh Press, 2003, pp. 64–88.

U.S. Census Bureau, "QuickFacts: Allegheny County, Pennsylvania," last revised 2015. As of March 20, 2017:
http://www.census.gov/quickfacts/table/PST045215/42003,00

U.S. Code, Title 33, Navigation and Navigable Waters, Section 1251, Congressional Declaration of Goals and Policy, 1972.

U.S. Environmental Protection Agency, "What Is Green Infrastructure?" undated. As of March 20, 2017:
https://www.epa.gov/green-infrastructure/what-green-infrastructure

———, Office of Wastewater Management, *Combined Sewer Overflows: Guidance for Long-Term Control Plan*, Washington, D.C., EPA 832-B-95-002, January 1995.

———, Office of Wastewater Management, *Combined Sewer Overflows: Guidance for Monitoring and Modeling*, Washington, D.C., EPA 832-B-99-002, January 1999.

———, *United States' Notice of Lodging of Proposed Consent Decree and Motion for Stay of Litigation*, Washington, D.C.: U.S. Environmental Protection Agency Environmental Enforcement Section, 2007. As of March 20, 2017:
https://www.epa.gov/sites/production/files/2013-09/documents/alcosan-cd.pdf

———, *A Screening Assessment of the Potential Impacts of Climate Change on Combined Sewer Overflow Mitigation in the Great Lakes and New England Regions*, Washington, D.C.: National Center for Environment Assessment, Office of Research and Development, EPA/600/R-07/033F, February 2008. As of March 20, 2017:
https://cfpub.epa.gov/si/si_public_record_Report.cfm?dirEntryId=188306&CFID=6733197&CFTOKEN=72223449&jsessionid=5a3025b141a0b1f861001a2f744521787568

———, *Greening CSO Plans: Planning and Modeling Green Infrastructure for Combined Sewer Overflow (CSO) Control*, Washington, D.C., Publication No. 832-R-14-001, March 2014. As of March 20, 2017:
https://www.epa.gov/sites/production/files/2015-10/documents/greening_cso_plans_0.pdf

USEPA—*See* U.S. Environmental Protection Agency.

Valderrama, A., L. Levine, E. Bloomgarden, R. Bayon, K. Wachowicz, and C. Kaiser, *Creating Clean Water Cash Flows: Developing Private Markets for Stormwater Infrastructure in Philadelphia*, Arlington, Va.: Natural Resources Defense Council, EKO Asset Management Partners, Nature Conservancy, Issue Brief IB:13-02-A, February 2013. As of March 20, 2017:
http://www.nature.org/ourinitiatives/regions/northamerica/unitedstates/pennsylvania/pa-issue-brief.pdf

van Oldenborgh, G., F. Doblas Reyes, S. Drijfhout, and E. Hawkins, "Reliability of Regional Climate Model Trends," *Environmental Research Letters*, Vol. 8, No. 1, 2013.

Volkening, Aaron, photo of bioswale, Flickr, 2010.

Walker, W., M. Haasnoot, and J. Kwakkel, "Adapt or Perish: A Review of Planning Approaches for Adaptation under Deep Uncertainty," *Sustainability*, Vol. 5, No. 3, 2013, pp. 955–979.

Water Environment Federation, "The Real Cost of Green Infrastructure," December 2, 2015. As of March 17, 2017:
http://stormwater.wef.org/2015/12/real-cost-green-infrastructure/

Weaver, C., R. J. Lempert, C. Brown, J. Hall, D. Revell, and D. Sarewitz, "Improving the Contribution of Climate Model Information to Decision Making: The Value and Demands of Robust Decision Frameworks," *WIREs Climate Change*, Vol. 4, No. 1, 2013, pp. 39–60.

Wilks, D., and R. Wilby, "The Weather Generation Game: A Review of Stochastic Weather Models," *Progress in Physical Geography*, Vol. 23, No. 3, 1999, pp. 329–357.

Wood, A., L. Leung, V. Sridhar, and D. Lettenmaier, "Hydrologic Implications of Dynamical and Statistical Approaches to Downscaling Climate Model Outputs," *Climate Change*, Vol. 62, No. 1–3, 2004, pp. 189–216.

Woodward, M., Z. Kapelan, and B. Gouldby, "Adaptive Flood Risk Management Under Climate Change Uncertainty Using Real Options and Optimization," *Risk Analysis*, Vol. 34, No. 1, 2014, pp. 75–92.

Yosie, T. F., *Retrospective Analysis of Water Supply and Wastewater Policies in Pittsburgh, 1800–1959*, Carnegie Mellon University, doctor of arts dissertation, 1981.